FOREVER ROSE

150 YEARS OF
STEM EDUCATION

Forever Rose: 150 Years of STEM Education
Published by Rose-Hulman Institute of Technology
Terre Haute, Indiana, U.S.A.

Library of Congress Control Number: 2024916005

ROSE-HULMAN INSTITUTE OF TECHNOLOGY, Author & Publisher
FOREVER ROSE

ISBN: 979-8-218-48427-9 (paperback)
ISBN: 979-8-3304-5055-8 (digital)

EDUCATION / Organizations & Institutions
EDUCATION / History
EDUCATION / Teaching / Subjects / Science & Technology

Content curated by Rose-Hulman's Office of Communications and Marketing, with special acknowledgment to Dale Long, Shaun Hussey, and Paul Shepherd.

Book Design: Michelle M. White (mmwbooks.com)
Publishing Management: Susie Schaefer (finishthebookpublishing.com)

QUANTITY PURCHASES:
Schools, companies, professional groups, clubs, and other organizations may qualify for special terms when ordering quantities of this title.
For information, email communications@rose-hulman.edu

Introduction

For 150 years, Rose-Hulman has been a beacon of inspiration, innovation, and community. Since its inception in 1874, the tapestry of our history is woven with countless stories that embody the spirit and achievements of our exceptional community. From the cornerstone moments of our founding to the vibrant celebrations of our sesquicentennial, each of the following vignettes serves as a window into the enduring legacy of Rose-Hulman and the remarkable individuals who have shaped it.

It all began in 1874, with Chauncey Rose's visionary dream of an institution dedicated to the practical education of young men in engineering, science, and mathematics becoming a reality. Rose Polytechnic Institute, as it would become known, laid the foundation for a tradition of excellence that has flourished for a century and a half. Our first president, Charles O. Thompson, set the standard for academic rigor and leadership that continues to inspire our community.

Throughout the decades, figures such as Herman Moench, whose dedication to teaching and the engineering profession left an indelible mark on the institution, and Anton Hulman, whose generosity and support helped propel Rose-Hulman into a new era of growth and innovation, have been central to our story. Samuel F. Hulbert's tenure as president saw unprecedented advancements, helping to shape the Rose-Hulman of today. President Robert A. Coons' 35-plus years of experience at Rose is now laying the foundation for the next 150 years and beyond.

Milestones such as the establishment of our first graduate programs, the construction of cutting-edge facilities, committing to coeducation in 1991, and our rise to national prominence in undergraduate engineering education are testaments to our unwavering commitment to excellence. Our sesquicentennial celebrations not only honor these achievements but also reflect the vibrant and dynamic community that defines Rose-Hulman.

As we commemorate 150 years of history, we celebrate the countless contributions of our alumni, faculty, staff, and students. Each story captured in this book is a testament to the enduring spirit of Rose-Hulman, a spirit that continues to inspire and innovate. This collection of milestones serves as a tribute to our past, a reflection of our present, and a beacon lighting the way to our future.

Chauncey Rose and nine of his friends established the Terre Haute School of Industrial Science "for the intellectual and practical education of young men."

September 10, 1874

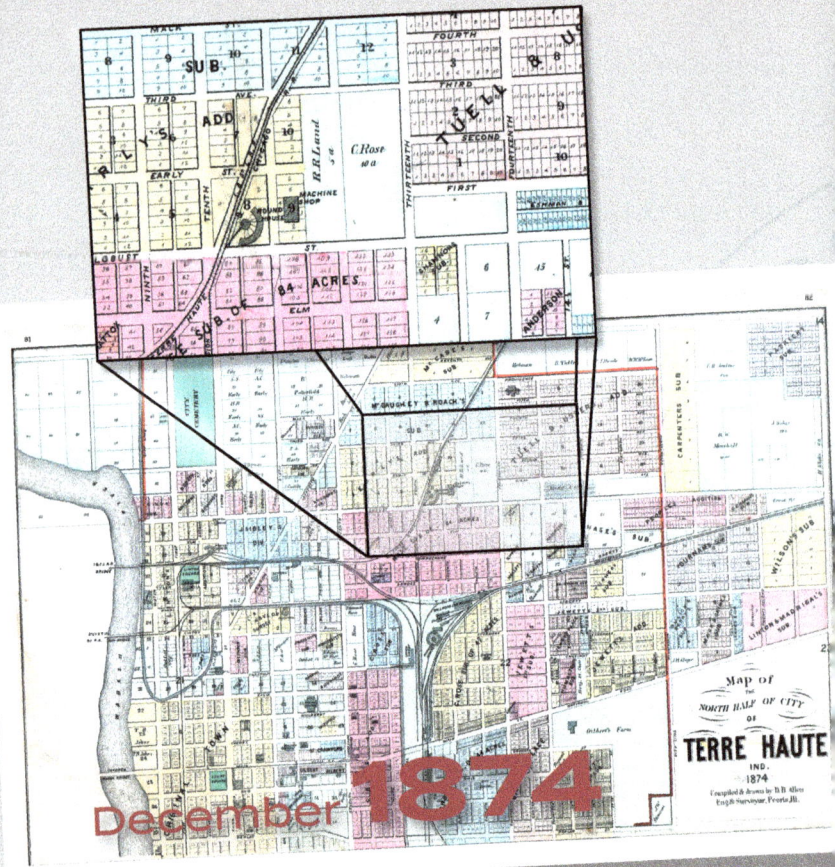

December 1874

Map of the NORTH HALF OF CITY of TERRE HAUTE IND. 1874

Chauncey Rose provided 10 acres of land for building a new college, which had been incorporated as the Terre Haute School of Industrial Science. This location is now at the corner of 13th and Locust streets in downtown Terre Haute. A contract was later awarded to construct an academic building for $81,000, with heating added in January 1876 for $4,000, and extra amenities to campus costing an additional $1,700.

The cornerstone of the college's original academic building (four-floor structure with 46 rooms) was laid at the corner of 13th and Locust streets in Terre Haute. The school's name was changed to Rose Polytechnic Institute.

September 11, **1875**

Chauncey Rose died in Terre Haute at age 88. His will provided for a bequest of more than $100,000, bringing the value of his total donations to more than $500,000. This would be approximately $14 million today.

August 13, 1877

1883

Mechanical engineering was the only major offered when Rose Polytechnic Institute opened.

March 7, **1883**

EXCERPTS FROM

THE INAUGURAL ADDRESS OF
DR. CHARLES O. THOMPSON

FIRST PRESIDENT OF

ROSE POLYTECHNIC INSTITUTE

"ENGINEERING IS THE TERM THAT INCLUDES ALL THE ARTS OF PRODUCTION AND CONSTRUCTION WHICH ARISE FROM THE PHYSICAL SCIENCES. ITS OBJECT IS TO BEND THE FORCES OF NATURE TO THE SERVICE OF MAN."

". . . THE ENGINEER IS DISTINCT FROM THE ARTISAN OR CRAFTSMAN (TECHNICIAN) BY EXACTLY THE AMOUNT OF HIS KNOWLEDGE OF THE SCIENTIFIC PRINCIPLES WHICH UNDERLIE THE PRACTICE OF HIS PROFESSION AND HIS RESULTING ABILITY TO APPLY THOSE PRINCIPLES TO THE READY AND COMPLETE SOLUTION OF REAL PROBLEMS AS THEY ARISE."

"NO GRADUATE OF ANY SCHOOL IS . . . AN ENGINEER. THE QUALITIES OF GOOD JUDGMENT AND EFFICIENT REASON GROW ONLY IN THE ATMOSPHERE OF EXPERIENCE."

MARCH 7, 1883

The Institute opened with exercises for inaugural President Charles O. Thompson, PhD. Twenty-five students enrolled with 16 freshmen, 26 sophomores, and three juniors. All were mechanical engineering majors.

Three graduates earned the college's first diplomas. The Commencement also served as a memorial service for founding President Charles O. Thompson, PhD, who died on March 17, 1885 from a sudden illness.

Charles O. Thompson

June 25, 1885

THE INDIANAPOLIS JOURNAL.

Rose Polytechnic and DePauw University Turn Out a Grist of Students.

First Class of Graduates from Terre Haute's Mechanical School—Important Changes in DePauw Faculty and Government.

ROSE POLYTECHNIC.

Graduation of the First Class in This School of Mechanical and Industrial Work.

Special to the Indianapolis Journal.

TERRE HAUTE, June 25.—The commencement exercises at the Rose Polytechnic, to-day, marks an epoch in the history of an institution the only one of its kind in the West, and only paralleled in its scope by its Eastern prototype—the Worcester Free Institute. The school, as is well known, was conceived and endowed by Terre Haute's most liberal citizen, whose name it bears, and whose memory it perpetuates—Chauncey Rose, who furnished it with an endowment fund of $400,000, and before his death selected the grounds and erected the college building and machine shops.

Mr. Rose has also remembered Vigo county in the Terre Haute Orphan's Home, where, from an endowment of $300,000, interest amounting to

1887

The first civil engineering degrees were offered.

Malverd A. Howe

Professor of Civil Engineering

The first Heminway Gold Medal was presented at Commencement to the graduating senior having the highest class standing. Sarah Heminway donated $1,000 to endow the annual award, along with a bronze medal to the student having the highest standing after completing the first year.

Spring **1888**

Rose Polytechnic Institute became the first American college west of the Allegheny Mountains to award a chemical engineering degree, presented to Walter B. Riley.

Spring 1889

THE CHEMICAL LABORATORY AND SHOPS.

1893

Rose Polytechnic Institute started an
electrical engineering degree program.

Greek organizations started to become a part of campus life, with Alpha Tau Omega becoming the first national social fraternity in the fall of 1893. Rose fraternities have earned national distinction for their community service, philanthropy, academic achievements, and campus involvement. Chi Omega became the campus' first sorority in the spring of 1996.

November **1893**

C. H. FRY, JR.   J. D. INGLE, JR.   H. T. LEGGETT.   F G HUNT.   E. L. SHANEBERGER.   J. T. MONTGOMERY.
W. R. SANBORN.   W. S. SPEED.   L. E TROXLER   W. O. MUNDY.   W. I. DECKER.
G. WILLIUS, JR.   F. E. SMITH, JR.   F. F. SINKS

The legend of Rosie began after students searched the countryside around the college in Terre Haute for useful elephant images to use as a mascot, inspired by the elephant's reputation for having a long memory. In one memorable case in 1902, recorded by an Indianapolis newspaper, two students were arrested for stealing a large elephant sign that was brought to campus and used as an athletics scoreboard. An elephant mascot, eventually named "Rosie," made its first appearance at a 1911 baseball game between Rose Poly and Indiana State Normal School (now Indiana State University).

1908

Rose Polytechnic earned Carnegie Foundation accreditation for being of "College Rank." Serious discussions began on moving the college to provide space for growth. Site possibilities were explored throughout the city.

The Board of Managers, faculty, and alumni club representatives visited the Hulman Farm, a 123-acre tract that was five miles east of downtown. Three buildings were proposed, costing approximately $250,000. The land was donated from Herman and Anton Hulman Sr.

June 1914

Fall **1915**

The first fundraising campaign commenced with a $500,000 goal to provide the first $300,000 for buildings (to be raised within three years) and $200,000 for an endowment fund (within five years).

Demas Deming donated $100,000, which helped pave the way for construction of the first residence hall, later named Deming Hall, on the new campus.

DEMING MEMORIAL DORMITORY

1921

The Institute's "Dear Old Rose" was written by 1922 chemical engineering alum Malcolm Scott. He played in big bands in New York City before returning to Terre Haute, and later added a degree in music from Indiana State Teacher's College (now Indiana State University). Music and lyrics for the Institute's Alma Mater were composed by ISU music professor Raymond Mech and later updated by his son, Andrew Mech, PhD, a Rose-Hulman emeritus professor of mechanical engineering.

September 13, 1922

The cornerstone of the Main Academic Building (now known as Moench Hall) was laid, and the college was open for classes, even though construction was not entirely completed.

All work on the original Main Building was completed.
A gymnasium was featured at the north end of the top floor.

January **1923**

Deming Hall opened as the Institute's first residence hall, as well as being a student center, dining area, and having several small classrooms. Construction cost was $100,000.

August **1926**

Community residents came to campus to view the first Rose Show scientific open house. It would become a popular annual event, attracting hundreds of guests to view Rose Poly's version of the World's Fair.

Spring 1928

Fall **1928**

Phil Brown became coach of all varsity sports but had greatest success on the gridiron, where his teams had a 99–109–7 record in 31 seasons. He coached the only two undefeated seasons in Rose football history (7–0 in 1941/8–0 in 1958). He was named an All-American Coach in 1930 by the *Cincinnati Enquirer*, and was a 1985 Indiana Football Hall of Fame inductee. The Rose football field was named in Brown's honor in 1969.

1931

CLASSES

A graduate program was started in 1931, even though records show that the first master's degree from Rose Polytechnic Institute was conferred in 1892 to Taro Tsuji, the son of a member of Japan's legislature.

Wilbur Shook, a 1911 alumnus, donated a surplus B-29 airplane hangar that became a fieldhouse for athletics and student recreation on campus. Construction cost was $310,275. The building was later named Shook Fieldhouse in 1961 to honor his many contributions to the Institute.

Fall 1948

SHOOK, WILBUR BRYANT.

Born May 11, 1889, at Versailles, Indiana. Graduated from Terre Haute High School. Artist on Modulus Staff. Foot Ball '09, Base Ball '08-'09-'10. Athletic Director '09-'10. V. Q. V Fraternity. Course: Architecture. Home Address: Terre Haute, Ind.

"Hard features every bungler can command;
To draw true beauty shows a master's hand."

March 25, **1953**

The Oscar C. Schmidt Memorial Lecture Series began
to bring prominent leaders in business and industry for a
campus convocation. Schmidt had been president of the
Cincinnati Butchers Supply Company. The series also
was supported by Schmidt's son, C. Oscar Schmidt.

The campus' first formal student center was opened in 1954 with spaces for students to gather and relax between classes. It later became the Templeton Administration Building in honor of Robert J. Templeton, which housed Offices for the President, Registrar, Development, Career Services, and Alumni Relations, along with the campus print shop. The original building cost was $51,425, plus $10,500 for furnishings. Renovations in 1968 cost $148,000.

1956

The opening of the Baur-Sames-Bogart Hall brought more students to live on campus. The building cost was $356,680. The building was named in honor of Oscar Baur (Class of 1887), Charles Sames (Class of 1886), and Paul N. Bogart (a former trustee).

Degrees in chemistry, mathematics, and physics were added to the curriculum.

1958

Fall **1960**

The Institute purchased its first computer under the leadership of Darrell Criss, PhD, a longtime professor who would become head of the Department of Computer Science. This technology provided a foundation for the bachelor's degree program in computer science to be added to the curriculum in 1968.

The Lynn H. Reeder astronomical laboratory and observatory
was added to the west end of campus.

Speed Hall opened to meet the growing demand for student campus housing. Eighty-five students moved in to start the second semester. The building was named in honor of 1895 alumnus William S. Speed. Construction costs were $535,000.

Fall 1964

The baseball field was named in honor of 1914 electrical engineering alumnus Art Nehf, who had been a highly decorated Major League Baseball pitcher, with a 184–120 career record in 15 major league seasons. Nehf was a member of New York Giants teams that won the World Series in 1921 and 1922, and was the winning pitcher in the series-winning games both seasons. He also played in the World Series in 1923 and 1924, and pitched for the Chicago Cubs in the 1929 World Series. Nehf led the National League in complete games in 1918, including a 21-inning performance on August 1. He is a member of the Rose-Hulman and the Indiana Baseball Halls of Fame.

Nehf, Pitcher

ROBERT M. ARTHUR
Associate Professor of
Bioengineering

BIOLOGICAL ENGINEERING

Biological Engineering is a combination of engineering and biology so that both can be fully utilized for the benefit of man. Graduates of this department, with their broad education in biological and engineering sciences can be expected to work in any area of technology which combines engineering and biology. These include: Medical Engineering, Environmental Health Engineering, Bionics, and Human Factors Engineering.

FRANK FREEDMAN
Associate Professor
of Bioengineering

Hürsta, William N.

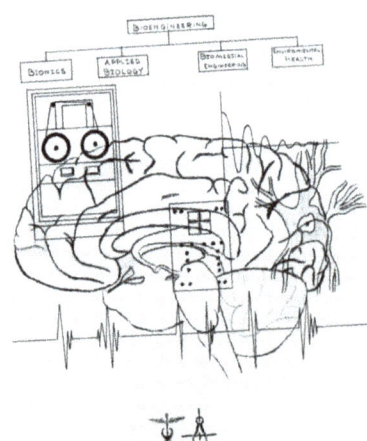

1966

Photo from 1968 Modulus

A degree in biological engineering was added to the curriculum.

The first group of rising high school seniors attended the pilot year of the new Operation Catapult on-campus summer STEM camp. Sixty-seven students participated (higher than the original 40-student goal). The project-oriented program allows students to study and do research in a STEM area of interest while also getting a look at college life. Still active in 2024, Catapult was founded by Director of Admissions Paul Headdy and Professor of Mathematics/1949 alumnus Alfred Schmidt. In the years since, many of the Catapulters have returned to earn degrees from Rose-Hulman and launch successful STEM careers.

June **1967**

1968

Computer science was added as a degree within the Department of Mathematics.

The Triplet residence halls — Blumberg, Mees, and Scharpenburg — were constructed to welcome more students living on campus. The cost of the buildings was $1,021,650.

August 1968

A 150,000-gallon elevated water tank was constructed to provide much better fire protection for campus buildings.

Fall 1969

October
1970

Crapo Hall was completed to provide additional classrooms for
the growing academic programs and was opened for classes
in December. The building was named in honor of alumnus/
trustee Frederick Crapo and his wife, Mildred.

December 1970

Anton and Mary Hulman transferred all Hulman Foundation assets to the Institute. The college's name officially changed to Rose-Hulman Institute of Technology in January 1971. The Internal Revenue Service set the value of the Foundation at $11,000,000, but the actual value was probably more. The increased financial resources allowed for an increase in the faculty in many academic areas.

A degree program in aerospace engineering was dropped, and biological engineering was replaced by an environmental engineering program that became an option in the Department of Civil Engineering. Aerospace courses would still be offered as electives in the curriculum.

September
1973

Prof. Rex Robinson gives his class of civil engineers an exercise in taking notes.

Fall **1974**

Construction of the New Learning Resource Center building was completed in the summer of 1974. It was renamed the John A. Logan Library when it opened that fall. Prior to its construction, the library had been a large, two-story room with metal bookshelves, reading tables, a card catalog, and a checkout desk in the B-section of Moench Hall. A learning center was added to the space in the late 1970s.

The Main Building was named Moench Hall in honor of the contributions of 1929 electrical engineering alumnus Herman A. Moench, who spent 56 years as professor, Department of Electrical Engineering chair, acting president, vice president, and senior vice president. The Moench Distinguished Senior Commendation and Moench Distinguished Professorship were named after him.

A promotional poster featuring a Rose-Hulman student skiing down the "slopes" of a snow-covered Indiana cornfield in front of an old barn was created to introduce Rose-Hulman to prospective students throughout the country. The backside of the Ski Terre Haute poster contained a variety of information, much of it humorous, about Rose-Hulman, Terre Haute, and Indiana. Ski Terre Haute posters have become cherished by alumni. The campaign was organized by Director of Admissions Duncan Murdoch, Assistant Director Chuck Howard, Vice President of Business Affairs Tom Mason, Athletic Director John Mutchner, and others.

Fall,
1980

Deming and Baur-Sames-Bogart residence halls were remodeled to add housing for an additional 45 students to meet increased enrollment demand, which was a record 1,268 students enrolled.

A $4.75 million grant from the Olin Foundation supported the building of a new classroom/laboratory building, later named Olin Hall (dedicated on October 21–22, 1983). An additional fundraising campaign supported the addition of Hadley Hall (named in honor of benefactor/1929 alumnus George Hadley and his wife, Mary, creator of Louisville-based Hadley Pottery Company, after providing $1.25 million from their estate) for administrative and admissions offices that provided a link between Olin and Moench Halls. The combined cost of the two buildings was $6.2 million.

July 1981

the Rose Thorn

Vol. 19, No. 8   Rose-Hulman Institute of Technology   OCTOBER 21, 1983

Olin Dedication Saturday

Formal dedication ceremonies for Olin Hall, a $4.75 million new academic building, are scheduled for 10 a.m., Saturday, Oct. 22.

The building, which houses classrooms and laboratories for chemical engineering, civil engineering, and the life sciences, was underwritten by the Olin Foundation. The foundation funded the design, construction, fixed equipment and furnishings for the 52,000-square-foot structure.

Four trustees of the Olin Foundation will arrive on campus Friday afternoon for an informal tour of the building and an opportunity to meet with students and faculty members who work and study in these new facilities. A dinner honoring the Olin Foundation trustees is scheduled later in the evening.

The Olin Foundation was established in 1938 by the late

colleges and universities. Since its inception, its grants for this purpose have exceeded $100 million.

The Olin Foundation has been committed to quality education since its founding and has provided excellent facilities at many of the nation's most prestigious institutions. President Holbert said. "The announcement in September, 1981 that Rose-Hulman Institute of Technology had been selected for an Olin-funded building was tantamount to receiving the 'Good Housekeeping Seal of Approval' for quality higher education.

Designed by VOA Associates, Inc., Chicago-based architects who have developed Rose-Hulman's long-range campus master plan, Olin Hall complements the 61-year-old Moench Hall and is an integral part of the newly-created focal point and

Olin Hall will be dedicated in ceremonies this weekend. See pages 4 and 5 for details.

Fall **1982**

A five-week intensive course, Fast Track Calculus, began for select incoming first-year students.

The Institute's computer commission, part of an earlier "To the Beat of a Different Drummer" study, both organized by Vice President for Planning and Data Systems A. Thomas Roper, recommended implementing computers into all aspects of the college. This included providing personal desktop computers to faculty, students, and office operations, and a complete review of all curricular matters to implement computerized instruction. Clusters of computers were available to students in public areas throughout campus.

April **1984**

Spring **1986**

Chapman Root and his wife, Susan, donated $500,000 toward the Moench Hall renovation project. The Root Quadrangle was dedicated in their honor on October 8, 1986.

Spring **1986**

David Against Goliath

The baseball team captured the Mayor's Cup with a victory against crosstown rival Indiana State University at Art Nehf Field. The win became the topic of national media attention, as several members of the Rose team had to leave the game before its conclusion to take an exam on campus. Radio newscaster Paul Harvey stated, "First things come first at Rose-Hulman."

During a 17-day trip to Europe, Rose-Hulman became the first American college basketball team to play a game in the Soviet Union. The trip was organized by then-Athletic Director and Head Coach John Mutchner. The team experienced some travel mishaps along the journey. National media reported that Mutchner was forced to make substitutions by shoe size during games.

FRIDAY, DEC. 11, 1987

ROSE THORN

Sports

PAGE 3

Front: Mark Christman, Dustin DeHaven, head coach John Mutchner, Ricky Meyer, Mike Webster and manager Mike Lindsay; middle: assistant coach Jeff Justus, Ron Steinhart, Brett Falhauer, Chad Rehmeyer, Jeff Harrison and assistant coach Jim Hargis; back: assistant coach Kelly Land, Britt Petty, Doug Underwood, Trevor Olsen, John Lacheta, David Urbanek and Rodney Adams. Not pictured is Tim Cindric.

Rose prepares for European trip

Rose-Hulman basketball players and coaches will spend part of their Christmas vacation away from their families this year. But they don't mind.

The team will be in Moscow playing an exhibition game as part of the fifth Engineer European trip.

This is the first time the Soviet Union has been included in the traditional European trip made every four years by the Rose-Hulman squad. It also is the first time an American college team has played a game in Russia. Every other American team playing in the USSR has been an all-star squad or a sponsored team such as Athletes in Action.

"According to ABAUSA (the amateur basketball governing body), Rose will be the first American college team to play in the USSR," Coach John Mutchner said. "That is a significant breakthrough for Rose and will provide an excellent educational opportunity for the players."

The team will leave Terre Haute on Dec. 16 for a 17-day trip that will take it to six countries.

The squad will be in Moscow on Dec. 19-22 and will also play in France, England, Germany, Belgium and the Netherlands. The team will return to Terre Haute on Jan. 1.

Mutchner said who the Engineers will play in their seven games abroad is not that important.

"The educational opportunity for our student-athletes far outweighs any athletic goals we might have," he said.

"Our objectives are to enable the players to have a great educational experience, to have fun and then to worry about winning basketball games," he added.

Trips to Europe and places such as Hawaii, Mexico, the Bahamas and Canada have become a tradition at Rose-Hulman during Mutchner's 18-year career here.

"One of the things I'm most proud of is that every player since 1971 who has been a member of our basketball team for four years has had an opportunity to visit Europe," Mutchner pointed out.

Spring **1987**

Educators from throughout the world came to campus and Terre Haute to attend the 17th Frontiers in Education conference, discussing issues regarding the future of engineering and science education.

Fall **1987**

May 1988

An eight-foot-tall steel sculpture that vividly displays a variety of steel shapes and connections was placed in a plaza north of Olin Hall to commemorate the contributions of former civil engineering professor Cecil T. Lobo, PhD. He was a faculty member for 34 years and advisor of the student chapter of the American Society of Civil Engineers.

Degree programs in computer engineering and applied optics were added. Additional minors were also added in Spanish and Latin American studies.

Fall
1988

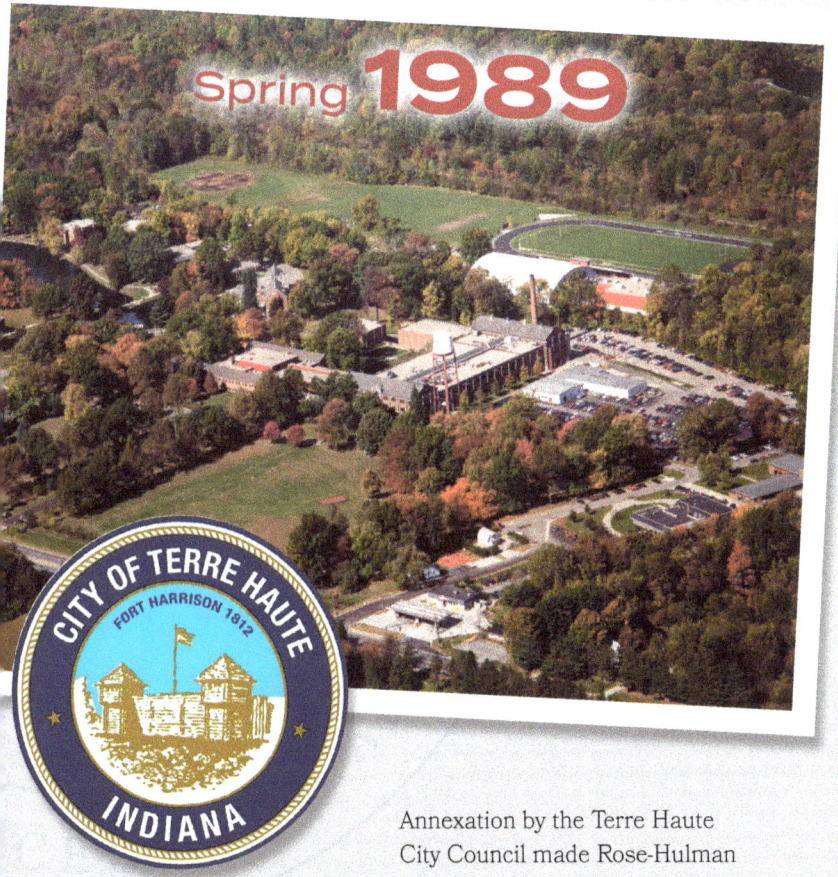

Spring **1989**

CITY OF TERRE HAUTE
FORT HARRISON 1812
INDIANA

Annexation by the Terre Haute City Council made Rose-Hulman officially a part of Terre Haute's city limits.

the Rose Thorn

Board approves mission

by Clark Penlice

On Oct. 5, Rose-Hulman's Board of Managers set out a mission to "establish Rose-Hulman as THE BEST undergraduate engineering and science college," by the year 2000.

Understanding the magnitude of the project, Rose-Hulman President Samuel F. Hulbert estimates the need to raise $80 to 200 million dollars over a ten year period.

The $80-90 million called for $90 million dollars to be raised.

The actual design will come from the Board of Managers,

Hulbert said that all information gathering should be complete by the end of January and that a first draft of the committee reports should be ready by spring.

Dr. Hulbert believes the Boards hands on involvement will keep them in tune with the happenings on campus and being in strong support of the Board.

The mission success becoming a co-ed institution.

According to Dr. Hulbert, "We cannot be THE BEST science and engineering school and be all male."

An emphasis will be given to the computer center, with up

year going to the upgrading of the facilities.

Student life will also be a center of focus. The construction of a new residence hall is still on the drawing board. An enlargement of the student life staff as well as counseling services were also mentioned.

The introductory physics laboratory equipment is also a concern and will be updated.

Greater attempts will be made to attract a diverse student and faculty population. Dr. Hulbert commented that very few science and engineering colleges are successful at attraction of

---

October
1989

Committee proposes Rose's mission for the '90s

EDITOR'S NOTE: At the time of publication, the Board of Managers had still not voted to approve the mission statement described in this article. In the next issue of the Rose Thorn, the results of the decision will be published.

by P.J. Hinton
News Editor

This week, Rose-Hulman's Board of Managers contemplation of Rose-Hulman's mission for the 1990s.

The mission, should they choose to accept, is to "establish Rose-Hulman as THE BEST undergraduate engineering and science college."

Emphasizing this statement, Rose-Hulman President Samuel F. Hulbert remarked at an administrative council meeting late last month, "We do not want to be great, nor good. Our goal is to make Rose-Hulman unquestionably the best."

How did this mission come to be? It is the result of an initial meeting of Rose-Hulman's strategic planning steering committee held last week. The meeting chaired by Mike Thomas, a Rose alumnus and a member of the Board of Managers, was held to determine what Hulbert described as "where we are now and the perceptions of what would be needed" for the coming years.

According to Hulbert, the

Mike Thomas

Samuel F. Hulbert

needed to propel Rose-Hulman into "a leadership role in science and engineering by the 21st century."

Members of this committee include Thomas, board members Clyde Willian, Tom Norman, Hal Brown, John Royse, John Titsworth, John Eagle, and President Hulbert. Also members are the Rose Vice Presidents James Eifert, Jess Lucas, Tom Mason, Ron Reeves, Tom Roper, and Chuck Howard.

Faculty members are Thad Smith, professor of political science, and Richard Ditteon, professor of physics and applied optics.

Admitting that being the best is "really a state of mind," Hulbert said that the committee as

used as criteria for accomplishing the mission.

The mission statement was submitted to the Board of Managers for approval at yesterday's meeting.

If the statement is approved, four subcommittees will go into action compiling and analyzing information to determine ways to meet this huge challenge.

The first committee, chaired by Eifert and Howard Freers, involves academic programs and facilities. They will be responsible for assessing personnel needs, equipment needs, "brick & mortar needs." In addition, they will determine whether any new degree programs should be considered in meeting its goals.

The second committee, concerned with student life and

facilities, is chaired by Jess Lucas and Clyde Willian. Problematic needs will be addressed by this committee. Hulbert realizes that there may be considerable overlap between the first and second committees' jobs. For example, a new auditorium would be useful in classroom facilities as well as for co-curricular activities like the dramatics club.

The third committee, chaired by Chuck Howard and Tom Norman, deals with financial aid. They will be responsible for setting goals to meet financial needs of the student population. The diversity of the student population will also be examined, especially in the recruitment of minorities and international students.

Ron Reeves and John Royse will chair the fourth committee, which will be addressing the issue of development and public relations. They will set goals to meet planning needs and determine ways to raise the amount of money to accomplish the mission.

The committees will meet every quarter Hulbert said that all information gathering should be complete by the end of January and that a first draft of each committee report should be ready by spring.

Once the four committees have completed their task, a fifth

investigate planning for the financial base of the mission. They will be faced with the task of figuring out the cost of the program, determining realistic objectives and setting priorities for the next decades.

Final reports from the five committees will be presented to the board on Oct. 5, 1990, if the mission is approved. At that time, a timetable for carrying out the mission will also be finalized.

Although the goals are still not well defined, there are some basic assumptions to the mission statement. According to Hulbert, Rose-Hulman will remain a predominantly undergraduate institution. Rose will also maintain approximately the same size.

The maider plan will "speak to the 'brick & mortar' needs." Computer facilities will be expanded, including the possibilities of a "wired campus" where all fraternity houses and residence halls will be connected to Rose computing facilities.

Most of all, it assumes that Rose will be a coeducational institution by the year 2000.

Other possibilities Hulbert mentioned included an enlarged student life staff as well as counselling services, enhanced laboratory facilities, expanded intramural and recreational facilities and more BSMLAB rooms in Crapo Hall (at least six

---

A revised Mission Statement by the Board of Managers set a goal for Rose-Hulman to provide the world's best undergraduate education in engineering and science by the year 2000. A long-range planning committee was established to examine future issues affecting areas of academics, admissions, student life, and development. Board Chairman John Titsworth stated, "Rose-Hulman wants to be the model for undergraduate engineering and science education ... We realize the challenge we have undertaken. It will take sizable resources and effective planning to accomplish our goal."

The innovative Integrated First Year Curriculum in Science, Engineering, and Mathematics (IFYCSEM) challenged a select group of students to recognize the relationships between calculus, chemistry, physics, computer programming, introduction to design, and graphical communications by taking those first-year courses together, instead of being taught separately. Central to the curriculum was the NeXT workstation, a new computer system developed by former Apple Computer co-founder Steve Jobs. National Science Foundation provided nearly $650,000 to support the groundbreaking academic initiative.

August 1990

Rose-Hulman established a global education program to provide select students with language skills and cultural awareness necessary for them to live and work in foreign countries. The Institute's East Asian Studies Program was among 11 U.S. colleges participating in National Science Foundation's Engineering Alliance for Global Education (EAGLE) initiative. The program started with Japanese language and culture courses; a partnership was opened between Rose-Hulman and China's Zhejiang University after a group of faculty visited campus. A group of Rose-Hulman faculty and administrators later visited the Chinese institution.

August **1990**

Spring **1991**

Rose-Hulman students answer homework hotline

by Will Mathies
Staff Reporter

The Rose-Hulman Learning Center and the local chapter of the National Society of Professional Engineers have teamed up to co-sponsor a unique community service. For the past year Rose students have been tutoring Vigo county students over the phone.

This "Homework Hotline" provides vital clues to students who need help in the areas of math and science. The hotline is available Sunday through Thursday from 7 pm to 10 pm at 877-8455. An average of 20 to 30 calls are made each week. While NSPE provides most of the manpower, not all of the tutors are members. So they can more easily answer questions the tutors have access to the same textbooks as the students. Typically, these questions cover algebra, trigonometry, calculus, high school chemistry and physics.

Currently the hotline only services the Vigo county area, but plans are underway to increase the service area to include some of the outlying areas. Some of these smaller communities do not have the resources of Terre Haute, making the homework hotline more valuable there. While the hotline is directed mainly at helping high school students,

some older night school students have called in seeking help Rose-Hulman's math and science background makes it a superb sponsor. Not only are the tutors interested and knowledgeable, the help they give promotes study in areas important to Rose-Hulman's future.

First starting out as a pilot program last spring, the hotline is busy finding ways to improve service. Recently, some 8,000 bookmarks were given to all the 6th through 12th grade students in Vigo county. Hopefully, these bookmarks will increase the number of calls by making the telephone number more available. Nightly radio service announcements are also being used to remind the High School students of the hotline.

The hotline is currently seeking a grant and a corporate sponsor to help Rose-Hulman cover the cost. Due to its unique nature, many other schools with strong technical background are interested in Rose's project. Susan Smith, the hotline coordinator, hopes to prepare a packaged nucleus that other schools could take to start their own programs. Additionally, Ms. Smith hopes to improve the tutor training by introducing the area teachers to the tutors so that they have a better idea of what the hotline is and what it can do for their students.

Dylan Schickel (right) and Andrij Petryna help local high school students with their homework.
photo by: Brian Dougherty, Staff Photographer

Students started providing free tutoring services to Vigo County and Clay County middle school and high school students through the Homework Hotline telephone service. It was in response to a Terre Haute Chamber of Commerce committee request to help improve the math/science skills of prospective employees. Financial assistance from the 3M Corporation supported toll-free telephone calls to come from Brazil, Indiana, and Hartford City, Indiana. Lilly Endowment Inc. support has allowed the service, now called AskRose, to continue to serve students throughout Indiana through a variety of tech tools.

Coeducation was on the horizon after the Board of Managers got the three-quarter majority vote necessary to allow female students to enroll, starting in the fall of 1995 (allowing all current students to graduate in an all-male environment). President Samuel F. Hulbert stated, "I believe that Rose-Hulman is headed for a rendezvous with greatness." The four-year delay allowed the administration, faculty, and staff to lay the groundwork on campus for a smooth transition.

October **1991**

Photos from 2003

Winter 1993

Rose receives part of $15 million NSF grant

Money to be used to improve engineering, science education

Rose-Hulman is one of six higher-education institutions chosen to participate in a $30 million, five-year, nationwide project to improve undergraduate engineering and science education.

The project, called the Foundation Coalition, will be funded by a $15 million matching grant from the National Science Foundation (NSF). Coalition members are required to match the NSF grant during a five-year period.

Other coalition members are Arizona State, Texas A&M (College Station), Texas A&M-Kingsville, and Texas Women's universities along with the University of Alabama. The coalition will work closely with Mesa and Glendale community colleges in Arizona.

The Coalition's goals are to reform undergraduate engineering and science education, attract and retain under-represented students to technical careers, and ease the transition of students entering four-year engineering or science degree programs.

The coalition's efforts will also focus on cooperative learning with student teams and between faculty to improve student problem-solving skills, according to Rose-Hulman electrical and computer engineering professor Jeff Froyd, coalition campus coordinator.

"Students improve their problem-solving abilities as they watch and learn, not only from faculty, but especially from other students," Froyd explained. "We want to improve how students tackle complex real-world problems," he said.

"At the same time, faculty will be collaborative learners, as they visit coalition campuses to exchange ideas about successful, new educational methods," Froyd stated.

A key goal of the coalition is to have a new Foundation Curriculum in place for the freshman and sophomore years at every coalition institution. The Foundation Curriculum will then be used to change junior/senior level classes to take advantage of new skills students have learned. The curriculum would be supported with the latest instructional technology including multimedia labs and classrooms equipped with computer workstations.

The differences among the coalition schools will be a strength in the group's efforts to create educational improvements, noted Gloria Rogers, dean for academic services at Rose-Hulman. "The schools are different in size, they recruit different types of students and they vary in their experience in educational reform," she said.

"As an example, The Texas Woman's University has successfully recruited female students into non-traditional areas of study. Its experience will be used to develop programs at Rose-Hulman and other coalition schools to increase female participation in engineering and science," explained Rogers, who will oversee the evaluation, assessment and dissemination of information about the coalition's efforts.

Froyd said some of the collaborative projects to be completed during the first year of the coalition's work include:

- The development at Rose-Hulman of a second-year integrated curriculum in specific academic areas. It will use elements of a curriculum developed by Texas A&M faculty.
- Arizona State will develop, in conjunction with Mesa and Glendale community colleges, courses modeled after Rose-Hulman's integrated first-year curriculum.

Rose-Hulman was invited to participate in the coalition because of its innovative integrated first-year curriculum and its leadership in using symbolic manipulation software to improve mathematics education. The integrated curriculum helps students understand the links between engineering, science and mathematics.

Coalition members have worked for almost a year to develop plans that will begin immediately to:

- Improve the first two years of engineering and science education by changing curricula at two-year and four-year institutions. Courses will be integrated to help students better understand the links between engineering, mathematics and the sciences; and the relationship between theory and practice.

- Develop programs and teaching methods to ease the transition of high-school and community college graduates into first-year and upper division engineering and science courses required to earn a B.S. degree.

- Create specific programs and teaching strategies to increase enrollment and retention of Hispanic, women and other under-represented minorities pursuing engineering and science degrees.

Rose-Hulman was selected to play a major role among seven U.S. colleges in a $15 million National Science Foundation grant to implement new ways to improve engineering education nationwide. The Institute received nearly $5 million. Major goals of the initiative were to improve the first two years of engineering and science education, implement modern instructional technology and techniques, develop programs and teaching methods to increase enrollment and retention of underrepresented minorities, emphasize cooperative learning, and ease the transition of students entering four-year engineering or science degree programs.

February
1993

A 410-member Commission on the Future of Rose-Hulman made its final report to the Board of Managers after 10,000 hours of meetings over 18 months by 10 task force committees. Commission members consisted of alumni (making up two-thirds of the group), business and industry representatives (200 different companies), education and foundation representatives, government officials, and faculty/staff members. Priority items became the guiding principles of the Vision to be the Best fundraising campaign. Areas of concentration included academics, admissions, student life, and development/fundraising.

A $2 million endowment, established with the assistance of a $500,000 challenge grant from the Kresge Foundation and additional support from the National Science Foundation, added at least $100,000 annually to the Institute's laboratory equipment purchases.

Spring 1993

Rose-Hulman established an educational partnership with Japan's Kanazawa Institute of Technology, one of the country's leading education institutions. Rose-Hulman students expanded their global horizons by learning Japanese on campus, and then taking a semester of language and culture courses in Japan. Faculty members have since taught at KIT, and basketball and baseball teams have played exhibition games against Japanese teams.

Fall
1993

September 1994

A cohort of 11 women students were housed at nearby Indiana State University while taking classes along with being enrolled in introductory classes and military science courses at Rose-Hulman. Eight students would eventually enter the college as sophomores and serve as Sophomore Advisors.

Fall 1994

The campus landscape grew by 69 acres resulting from the acquisition of land north and northeast of the campus. Work began to construct three large intramural fields on the property.

A 10-year, $100 million Vision to be the Best fundraising campaign was unveiled to increase funding for scholarships, expand campus computing systems, create a curriculum for the 21st century, construct new student service facilities, and enhance laboratory equipment. Ground was broken on a $10.9 million expansion of the Hulman Memorial Union, which brought new offices for student affairs and expanded career, health, and counseling services. After raising $116 million within the first four years, the goal was increased to $200 million.

March 1995

spring
1995

The Alfred R. Schmidt Bell Tower was added to the end of the front entranceway. The structure honored Schmidt, a 1949 alumnus who spent 46 years as a legendary mathematics professor and administrator.

Rose-Hulman unveiled the Institute's website, www.rose-hulman.edu, to keep alumni, prospective students, and others informed about what's happening on campus.

Spring 1995

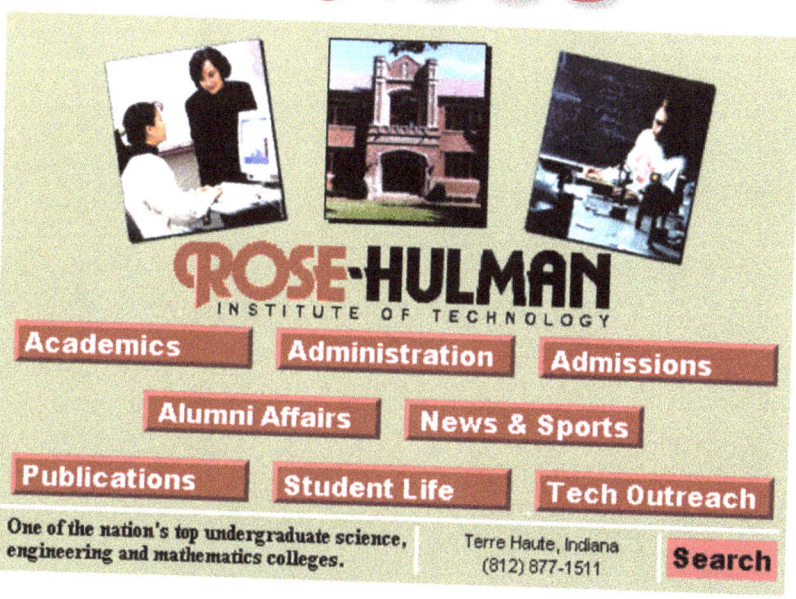

ROSE-HULMAN
INSTITUTE OF TECHNOLOGY

Academics   Administration   Admissions

Alumni Affairs   News & Sports

Publications   Student Life   Tech Outreach

One of the nation's top undergraduate science, engineering and mathematics colleges.

Terre Haute, Indiana
(812) 877-1511

Search

[Academics] [Administration] [Admissions]
[Alumni Affairs] [News]
[Publications] [Student Life] [Tech Outreach]
[Search]

The campus hosted the first overnight stop of Sunrayce 95, an Indianapolis-to-Golden, Colorado, solar car road race featuring Rose-Hulman's Solar Phantom and 39 other North American colleges. More than 3,000 people came to see the cars scattered across the front lawn and crossing the starting line along the front entranceway. Other spectators lined the race route throughout Terre Haute and western Vigo County. The scene was replicated in 1997 when the campus hosted the first overnight of Sunrayce 97 along the same route to Colorado Springs, Colorado. The Solar Phantom also participated in these cross country races in 1990, 1993, 1999 (finishing third), and 2001.

Sunrayce 95

AT ROSE-H
TERRE HAUT
JUNE

ROSE-HULMAN INSTITUTE OF TECHNO

PHANTOM

June 1995

August **1995**

Classes began for the first coed class that included 80 women comprising roughly 17.3% of the total student body. Female transfer students, including eight members of the consortium serving as Resident Assistants and Sophomore Advisors, played a vital role in acclimating the first-year coeds to all aspects of academic and student life.

August **1995**

A $3.2 million Olin Foundation Inc. gift supported expansion of Olin Hall to create eight advanced technology classrooms (now known as the Olin Advanced Learning Center). The building opened in fall 1997. The donation achieved a goal by the Commission on the Future of Rose-Hulman to provide flexible learning classrooms that support STEM education.

Rose-Hulman became one of the first undergraduate STEM colleges to require students to purchase laptop computers from the Institute, making it easier for the college to keep leading-edge technology in students' hands and benefit from the continued development of innovative computer-assisted instruction and new computerized classrooms and laboratories. "Perhaps no one tool (the laptop computer) has done more to increase student and faculty productivity," said Tom Roper, Vice President for Planning and Data Systems.

Fall **1995**

Spring **1996**

JOHN T. MYERS
CENTER FOR TECHNOLOGICAL RESEARCH
WITH INDUSTRY

A $4 million grant from Lilly Endowment Inc. created the Rose-Hulman Technology and Entrepreneurial Development Program that would develop a national model for project-based STEM education, create new laboratories for product/process development, increase the number of Indiana companies involved in project-based activities with Rose-Hulman, and create more internship and co-op programs for students with Indiana companies. This would establish the foundation for the eventual creation of the John T. Myers Center for Technological Research with Industry.

Beth (Knoy) Brock became Rose-Hulman's first female student to graduate with an undergraduate degree after completing 66 credit hours in three academic quarters. She is the daughter of 1958 alumnus/trustee Crone Knoy.

May
1996

Beth Knoy
Mechanical Engineering

Cook Stadium is dedicated and starts hosting football games, played on Phil Brown Field. The facility has concrete bleachers, a press box, a concession stand, and restrooms. It was named in honor of biomedical device pioneer William Alfred Cook for his support of the Institute.

October 1996

March 1997

The last sporting event in Shook Fieldhouse was an 86–69 victory by the men's basketball team against Washington University of St. Louis in the first round of the NCAA Division III national tournament. The Fightin' Engineers eventually suffered a last-second loss to Illinois Wesleyan University, a team that would go on to win the 1997 national championship.

Drawn by Rose-Hulman's growing national academic reputation, more than 350 engineering educators from throughout America and several foreign countries came to campus for the first Best Assessment Processes in Engineering Education Conference to learn about new standards for engineering accreditation. The symposium was supported by the Accreditation Board for Engineering and Technology and the National Science Foundation. A Rose-Hulman showcase featured several innovative academic initiatives. The conference was repeated with several hundred more engineering educators attending in 1998 and 2001.

Fall 1997

The 154,000-square-foot Sports and Recreation Center was opened to provide space for student recreation, athletics, and other campus and community events. The building featured the Institute's first indoor swimming pool.

Fall **1997**

In Washington, D.C., Rose-Hulman was one of the winners of the prestigious Theodore Hesburgh Award for exceptional faculty development programs designed to enhance undergraduate teaching and learning.

The Joy E. Hulbert Tennis Complex was dedicated. It is located between Cook Stadium and the east intramural/soccer field. It was named on behalf of the wife of former president Samuel F. Hulbert for her support and loyalty to the Institute and its students.

October 1998

Members of the inaugural female freshman class received their diplomas during Commencement. Chemistry/chemical engineering alumna Liza (Saunders) Davis received the Heminway Gold Medal for having the highest grade-point average in the class.

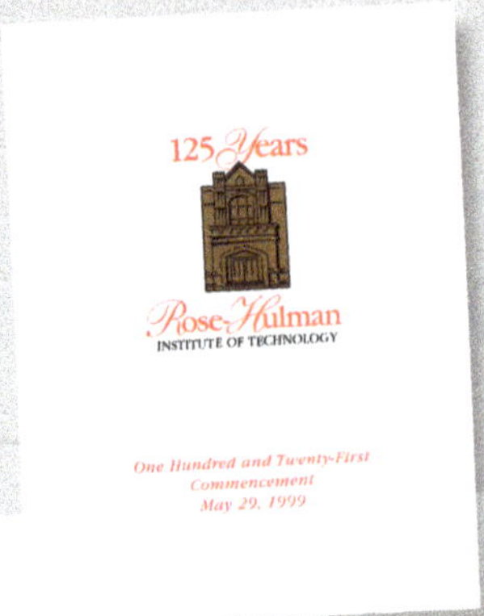

125 *Years*

Rose-Hulman
INSTITUTE OF TECHNOLOGY

One Hundred and Twenty-First Commencement
May 29, 1999

May **1999**

July
1999

The Indianapolis Colts brought their preseason training camp to Rose-Hulman's athletic facilities for the first of 11 summers, drawing crowds of fans and media attention to the college. During those 11 seasons, the team won one Super Bowl championship (2007) and played for another (2010).

Indianapolis Colts complete training camp

The Indianapolis Colts held their annual training camp at Rose-Hulman Institute of Technology for the first time this summer.

"It's a sad day," said Rose-Hulman President Samuel Hulbert when the squad left campus on Aug. 19. "The days have flown by. I'm going to miss the practices and hearing the stories of their days," said Hulbert.

The Colts considered their first camp at Rose-Hulman a successful venture.

"The facilities here were first class, and the people here treated us well," said quarterback Peyton Manning. "The food was good, which always helps. And my bed was comfortable. I thought it was a good change for us, and everything worked out well," said Manning.

"There's a lot more blue in Terre Haute today than there was a month ago," said Colts vice president of marketing and sales Ray Compton. "From a marketing perspective, it was a home run."

Admission and parking were both free to all fans throughout Colts Camp.

Rose-Hulman will serve as the host for the training camp for at least the next two summers.

Rose-Hulman was listed No. 1 among 132 specialized accredited engineering colleges in U.S. News & World Report's annual college guide — the first of 25 straight years of this distinction heading into 2024. "The ranking is a great testimonial to the quality of our people and programs. It is proof that Rose-Hulman has earned a national reputation as an educational leader," said President Samuel F. Hulbert.

August 30, **1999**

1874-1999 THE ROSE 125 YEARS

THORN

Volume 35, Issue 1   Rose-Hulman Institute of Technology   Terre Haute, Indiana   Friday, September 3, 1999

Rose-Hulman named "The Best"

Craig Pohlman
Editor-In-Chief

A big surprise was waiting for everyone as they arrived at Rose-Hulman this past week. Everyone could see the banner flying over the main entrance: "Rose-Hulman Ranked #1 by U.S. News and World Report."

We're on top according to deans and faculty of America's institutions that say we rank first in academic reputation among those schools not offering doctoral degrees. Rose-Hulman was above the rankings of colleges such as Harvey Mudd, Texas A&M, all military academies, and many California State institutions.

"I think it's great for the institution. It is symbolic of a lot of hard work, dedication, and persistence of students, faculty, and staff," says Thomas Miller, Assistant Dean of Students.

The news comes after the Solar Phantom team placed third in Sunrayce 99, the best finish in the school's history.

"At least now the parking stickers are true," says Junior mechanical engineer Karen Hill. "We are the best undergraduate science, mathematics, and engineering college."

For over the past two decades, Rose-Hulman President Dr. Samuel Hulbert has overseen the growth of the campus. The Vision to be the Best campaign has raised over $100 million to make improvements in the Rose-Hulman community. In only the past 5 years, the Hul-

A sign of everyone's hard work over the

man Memorial Union was remodeled and expanded, and new buildings such as the Olin Advanced Learning Center, Cook Stadium, Myers Building, and SRC were erected.

"You can't go without saying something about the great leadership under Dr. Hulbert for over the past twenty years. He has the vision moving us forward. The commitment of everyone has made Rose-Hulman what it is today," commented Miller.

In addition, the academic performance of entering Rose-Hulman students is impressive. According to Admissions of the 397

freshmen in the class of 2003, the mean SAT score was 1350, with 23 students receiving an 800 in math and 21 who has perfect verbal scores.

Rose-Hulman placed second in the previous rankings by U.S. News and World Report back in 1996.

According to Miller, "Rose-Hulman will continue to get positive exposure. When looking at engineering education, I believe that Rose-Hulman will be a household name like other similar institutions. We are something people will continue to talk about."

September 10–11,
1999

125 Years

Rose-Hulman
INSTITUTE OF TECHNOLOGY

1874-1999   THE ROSE

THORN

Volume 35, Issue 2   Rose-Hulman Institute of Technology   Terre Haute, Indiana   Friday, September 10, 1999

125 years strong
Rose-Hulman celebrates its day of founding

Rose-Hulman President Samuel Hulbert delivers his talk for the Last Lecture Series last spring. President Hulbert has no bone to pick about Rose-Hulman's growing success.

A 125th anniversary celebration featured a special convocation with the presentation of seven honorary degrees, a campus processional, and performances by the Terre Haute Symphony and Rose-Hulman Chorus. Officials announced the Institute received a record $48 million in gifts — the largest amount ever received in a year. Events included a speakers series and Young Master's Program, organized by students. A time capsule was placed behind the cornerstone of the new Sports and Recreation Center and a 50-year time capsule from beneath Shook Fieldhouse was opened at the celebration dinner.

September 1999

A $29.7 million Lilly Endowment Inc. grant created Rose-Hulman Ventures to provide technical expertise to help Indiana companies prosper, while providing faculty and students with cutting-edge educational and professional STEM opportunities. A follow-up $24.9 million grant provided by the Lilly Endowment in the spring of 2003 further expanded Rose-Hulman Ventures' operations and outreach.

The John T. Myers Center for Technological Research with Industry, now called Myers Hall, created a model for project-based engineering and science education programs. The 40,000-square-foot building provided workspace for competition teams, student projects, research projects, and collaborations with local business and industry. Supported by a $6.7 million grant from the U.S. Department of Energy, the building was named for the longtime district U.S. congressman.

October 1999

CTRI construction continues
Scheduled for completion in November

by Beth Bateman
Thorn News Editor

The John T. Myers Center for Technological Research with Industry (CTRI), which has been under construction since last year, is nearing completion. According to Wayne Spary, Vice President of Facilities and Operations, the building should be ready for occupancy by November 1.

The CTRI is a lab-style building which will house student projects. The L-shaped building is two stories tall with labs on both floors. The CTRI also contains office areas on the second floor, as well as a high bay lab, and an auditorium for presentations.

The labs are designed for a wide variety of flexibility so that they can be used for projects in many different majors, but some labs will have special features. For example, Spary explained that many of the labs on the lower level have isolated pads to accommodate projects which are sensitive to vibrations. The lower level labs also have no windows. Spary commented that this feature can be useful for light-sensitive projects, such as those done by applied optics majors.

The labs on the upper level rooms are all well-lit, with many windows throughout. Some of the upper level labs also have fume hoods which are useful for chemistry projects.

Although a great deal of work on the CTRI has been completed, there is much more still to be done. Workers are currently painting the interior and laying tiles in the bathrooms. Some drywall work is being completed, the floors are being laid, and glass for the building's windows will be installed soon. A bridge will also be built between the second level of the CTRI and the lower level of Moench Hall near the mailroom.

Although there is still more to be done, the work is expected to be finished in approximately six weeks. "It's all finish work now," commented Spary.

The CTRI building is still under construction, but is expected to be completed by early November.
Photo by John Ewoldt

A new 56,000-square-foot residence hall (dedicated as Percopo Hall in May 2005) provided additional space and services for 209 second-year students in hopes of improving retention in this important period in a Rose-Hulman student's academic development. A $5 million grant from the Lilly Endowment Inc. supported funding the new building and its operations. Total cost of the building was $11.3 million, which included funds to relocate the campus observatory. The building's name recognized the contributions of 1943 alumnus and trustee Michael Percopo and his wife, Christa.

November **1999**

April **2000**

The Oakley Observatory opened the universe to student exploration, with its multiple telescopes and other technological advances. The facility was named for Terre Haute's Oakley Foundation, which provided a $500,000 donation.

August 2001

New Applied Biology Degree: Supplying the future demand for biologist

Crystal Landreth
Staff Writer

Within the twenty-first century we will see many advances in drug discoveries, environmental management, gene therapy, and tissue engineering. It will be the biologists at the forefront of these new advances. Their research and discoveries will drastically impact our lives in the future.

To accommodate the future opportunities in biological-related occupations, Rose-Hulman has introduced the new Applied Biology Degree. What exactly is applied biology? The program will prepare graduates for professional careers in government and industrial research laboratories, and in the biotechnology and health-related industries.

Lee Waite, head of the Department of Applied Biology, noted that

Rose-Hulman students have always been successfully admitted to med school; however, the applied biology program will improve their preparation for a career in medicine.

Our health, our sustenance, and our environment are all based on biological processes. The importance of improvements in these subjects is obvious.

Rose-Hulman students are recognizing the necessities of biological

occupations. The school has witnessed a colossal increase in the number of students enrolled in biology courses. One hundred eighty two students were enrolled in biology courses during the past academic year compared to only 38 students two years ago. Eighteen students are currently enrolled as applied biology majors, eight females and ten males. The applied biology department was unable to recruit heavily last fall, due to the

uncertainty in the timeline for ing the AB B.S. degree; how with a stronger recruiting effc year, a large class of applied gists is expected in 2002.

Like all of Rose-Hulman' degrees, the applied biology is expected to be a great w producing the skilled biolog essary to impact the future.

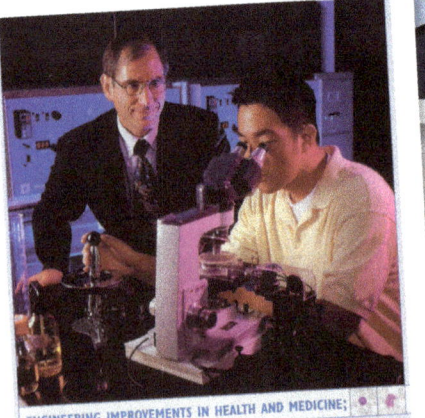

The new applied biology undergraduate degree program launched.

October
2001

The White Chapel brought added beauty to the west end of campus (overlooking Speed Lake) and has become one of the most picturesque campus locations, along with being a popular venue for weddings and special events for alumni and community residents. The building was named for 1947 alumnus John R. White and his wife, Elizabeth, who gave a $1.5 million gift to support the project. Other members of the White family have supported Rose-Hulman throughout the years.

Spring 2002

REACHING OUT

Rose-Hulman and Lilly Endowment working together to help educators and entrepreneurs throughout Indiana.

Thanks to almost $6.3 million from the Lilly Endowment of Indianapolis, Rose-Hulman will be expanding its service to students and teachers in Indiana middle schools and high schools, to entrepreneurs, and to the college's own students.

BY DAVID PIKER AND DALE LONG

The combination of these projects will allow Rose-Hulman to help create jobs, interest more middle- and high-school students in mathematics and science, expand access to ideas, give Rose-Hulman students more real-world project experience, and bring student-designed technical solutions to the marketplace.

This series of articles on pages 14-16 provides an in-depth look at the grant details and the impact they will have on Rose-Hulman and its outreach.

During the past eight months, the Endowment made the following gifts:
- $2.6 million to expand the Homework Hotline across Indiana school regions;
- $1.9 million to develop a Web Mall to improve chances of taking Rose-Hulman student ideas to the marketplace; and
- $1.77 million to create a Web portal to increase teaching resources available to middle-school teachers of science, mathematics and technology.

$2.6 Million Lilly Endowment Grant Paves Way For Homework Hotline Expansion Throughout Indiana

The telephones are ringing off the hook for Rose-Hulman Institute of Technology's Homework Hotline, giving tutors an opportunity to help middle- and high-school students to better comprehend the complexities of mathematics and science, and improve their problem-solving skills.

Through January of the 2001-2002 school year, the toll-free service helped 5,563 callers — well on the way to meeting the goal of 10,000 calls.

And, the prospects are even better in the future. A $2.6 million grant provided by the Lilly Endowment Inc. will help the Homework Hotline expand its reach across Indiana school regions, starting this fall.

The expansion will be done in three phases through the following schedule:
- 2002-2003: Northeast/East Central Indiana, including Fort Wayne, Anderson, Muncie and Richmond (Wayne County).
- 2003-2004: Southwest/Southeast Indiana districts, including Evansville, New Albany, Vincennes and Jeffersonville.
- 2004-2005: North Central/Northwestern Indiana districts, including South Bend, Kokomo, Lafayette and Gary/Hammond.

The $2.6 million grant continues a partnership that has paved the way for the Homework Hotline to become an important educational resource for students, teachers and parents, according to Learning Center/Homework Hotline Director Susan Smith.

The Endowment provided a $1 million grant in 1999 to expand the hotline into central Indiana school districts during the past two years. The service was started in 1990 after Rose-Hulman was approached by officials of the Greater Terre Haute Chamber of Commerce and Vigo County School Corporation; expanded into nearby Clay Community Schools (Clay County) in 1991; Blackford County (Hartford City) in 1992; and Monroe County and five west central school districts this school year.

"With the Homework Hotline, Rose-Hulman had the spark of an idea and the will to develop it," said Sara B. Cobb, Lilly Endowment vice president for education. "Now, through strategic collaborations involving Rose-Hulman, teachers, schools and students, Indiana middle- and high-school students who have questions about their math or science homework can get immediate help. The college students on the other end of the line don't just give them the answers, they teach them how to arrive at the correct results."

The Homework Hotline is available from 7 p.m. to 10 p.m. on Sundays through Thursdays during the school year. The toll-free telephone number is 1-877-ASK ROSE (1-877-275-7673). A total of 54 Rose-Hulman students serve as tutors, with 20 available to answer calls each night at the hotline's state-of-the-art communications center. ∎

Lilly Endowment Inc. provided $6.2 million in grants to support expansion of the Homework Hotline across Indiana school regions, develop a platform to improve chances of taking Rose-Hulman student ideas to the marketplace, and create a SMART web portal to increase teaching resources to middle school teachers of science, mathematics, and technology, which would eventually become the PRISM program.

A $400,000 grant from the W.M. Keck Foundation supported establishing a new microelectromechanical systems (MEMS) fabrication and application course and equipment laboratory in Moench Hall, supervised by the Department of Physics and Optical Engineering. Students can design and fabricate simple MEMS devices using silicon wafers.

Spring 2002

CAMPUS | News Notes

MICROELECTROMECHANICAL SYSTEMS TECHNOLOGY IS FOCUS OF NEW PROGRAM
MEMS technology creates microscopic, moving, electric devices

BY DAVID PIKER

A gear the diameter of a human hair and an optical mirror the size of the tip of a pin that can aim beams of light have been created using technology called microelectromechanical systems (MEMS), which is the focus of a new academic program at Rose-Hulman Institute of Technology.

A $400,000 grant from the W.M. Keck Foundation of Los Angeles has made it possible to develop a new MEMS fabrication and application course and to equip a new MEMS laboratory in Moench Hall. The new class was taught during spring quarter. Remodeling is currently under way in the B Section of the second floor of Moench Hall to change an existing area into a clean lab environment.

The W.M. Keck Foundation is the nation's largest philanthropic foundation that focuses its grantmaking primarily on the areas of medical research, science and engineering.

"MEMS technology creates microscopic, moving, electric devices so small that often gravity doesn't have an impact on their operation," explained Azad Siahmakoun, professor of physics and optical engineering, who is coordinating the introduction of MEMS at Rose-Hulman.

"MEMS will result in some of the most significant technical advancements of the 21st century," he stated "It is a technology that has been developed into a multi-billion dollar industry. The grant enables Rose-Hulman to begin educating students to meet the tremendous need for new engineering graduates who will have the expertise to further develop MEMS technology."

In a recent report, the Keck Foundation staff listed miniaturizing sensors, instruments, and computers as one of the five greatest opportunities to advance our knowledge and understanding of nature in the next decade.

Alumnus Chris Mack, who was among three persons that Rose-Hulman sought to review the preliminary proposal to seek funding for the MEMS initiative, says students will learn problem solving skills about unique microstructure problems that will be invaluable.

"Although direct MEMS experience will be valuable to some employers in the near future, more valuable will be the basic concepts of microstructure design and fabrication that students will learn," said Mack, vice president for technology at KLA-Tencor in Austin, Texas. The company is the world's leading supplier of process control and yield management solutions for the semiconductor and related microelectronics industries.

Because MEMS is having an impact on a wide range of professional fields, the new class is taught by a team of eight professors from five engineering and science departments.

"Students will learn that physics and chemistry influence the

Student Corey Trabaugh working in front of an oxidation furnace, inspects a silicon wafer upon which a MEMs device will be fabricated. The wafer is placed inside the furnace, and a thin oxide layer grows on the wafer surface. The layer then is patterned and etched using standard photolithographic techniques to produce a MEMS device.

mechanical systems can be developed for data acquisition and analysis," Siahmakoun said in explaining the interdisciplinary impact of the technology.

Chemists have borrowed technologies used in the fabrication of MEMS to develop miniaturized systems for chemical analysis, said Dan Morris, assistant professor of chemistry, who is one of the faculty teaching the new course. These systems are designed to carry out all aspects of chemical analysis, hence the birth of the micro-total analysis system or lab-on-a-chip.

"Performing chemical analyses on microchips offers the advantages of extremely small sample size requirements and superior separation of complicated samples in minimal amounts of time," Morris explained.

"Minimizing the amounts of materials and time required for chemical analysis is attractive, especially when one considers the importance of combinatorial chemistry in drug discovery and the explosion of biotechnology," he stated.

Students will design and fabricate simple MEMS devices using silicon wafers. Tiny mechanical devices with one or more moving parts can be fabricated on the silicon surface. "To enable students to understand such tiny motions, we will require them to design, model and characterize the electrical, mechanical or fluidic behavior of MEMS." Siahmakoun stated. A scanning electron microscope will be used to enable students to examine their device during and immediately after fabrication. ■

September
2002

The Flame of the Millenium Sculpture was placed on campus, surrounded by a fountain and brick walkway, after being moved from its original location near downtown Chicago. The 25-foot-tall stainless steel artwork, created by artist Leonardo Nierman of Mexico City, was donated to the Institute by 1969 alumnus Glen Raque and his wife, Barbara. The sculpture has become a feature of the Wabash Valley Art Spaces community art project.

October 2002

The William M. Welch Track and Field Complex was dedicated around Phil Brown Field at Cook Stadium. It was named in recognition of former track & field and cross country head coach Bill Welch, a member of the Indiana Track & Field Hall of Fame.

Hatfield Hall opened with a special concert featuring the Terre Haute Symphony Orchestra and student performing groups. The building has a 604-seat theater, alumni center, music classrooms/rehearsal rooms, Drama Club dressing and costume rooms, and offices for institutional advancement, communications and marketing, enrollment management, and alumni relations. The building was supported by 1984 alumnus and emeriti trustee Michael Hatfield and his wife, Deborah, and named on behalf of Mike's parents, Larry and Pat Hatfield.

The Department of Computer Science and Software Engineering moved into new offices, laboratories, and meeting/conference rooms in Moench Hall that formerly were used for an auditorium. The renovation project cost $1.8 million, which was covered by donations matched by Lilly Endowment Inc. The opening of Hatfield Hall's theater in 2002 freed up the space for this academic project.

September **2003**

New academic majors were added in biomedical engineering and software engineering, while a second major in biochemistry and molecular biology was also started. Academic departments were renamed for the Department of Applied Biology and Biomedical Engineering (ABBE) and Department of Computer Science and Software Engineering. The biochemistry and molecular biology major was part of the chemistry department. New laboratories in Myers and Moench halls were later created to support the new ABBE programs.

The 10-year Vision to be the Best fundraising campaign, the largest in Rose history at the time, far exceeded its original goal with $253 million in cash gifts and future commitments. The campaign's original goal was $100 million, but that increased twice after receiving widespread support from donors. The support brought $100 million in new facilities (including Olin Advanced Learning Center, John T. Myers Center for Technological Research with Industry, Sports and Recreation Center, White Chapel, Hatfield Hall, and expansion of the Hulman Union), $80 million in student financial aid, and $20 million for technology and laboratory equipment. Significant gifts included $14 million from 1984 alumnus Michael and Deborah Hatfield, a $7 million gift from 1943 alumnus Michael and Christa Percopo, and Lilly Endowment Inc. grants. It supported a number of new programs, including engineering physics and engineering management.

Vision to be the Best
ROSE-HULMAN INSTITUTE OF TECHNOLOGY

10–YEAR FUND–RAISING CAMPAIGN RECEIVES WIDESPREAD SUPPORT

June 30, 2004

Fall **2004**

The 75,000-square-foot Apartment East and West residence halls opened to provide a new type of on-campus living environment with apartment-style housing and additional meal options for 240 upper-class students. A commons area had a convenience store, a Subway restaurant, and a hair studio. The building cost $13.5 million.

A team of 75 students started working to re-engineer a 2005 Chevrolet Equinox in the U.S. Department of Energy and General Motors' Challenge X: Crossover to Sustainability Mobility national engineering competition. Fifteen select North American colleges were planning to reduce the compact SUV's energy consumption and emissions, while maintaining or exceeding the vehicle's performance.

Fall 2004

Challenge X team strives for success

Fred Webber
Staff Writer

At the beginning of the 2004-05 school year, Rose-Hulman began its participation in Challenge X. According to the Challenge X web site, Challenge X is a contest between 17 schools to transform a Chevy Equinox, a crossover sports utility vehicle, to minimize energy consumption, emissions, and greenhouse gases while maintaining or exceeding the vehicle's utility and performance. Rose-Hulman's Challenge X team is made up primarily of students from the Mechanical Engineering and Electrical and Computer Engineering departments, though a few people in Computer Science can participate. The project also has two faculty advisors, ME professor Zach Chambers and ECE professor Mark Herniter.

Dave Hoff, a senior chemical engineer, joined the Challenge X because of his interest in the automobile industry. "I've already worked two co-op rotations in industry and I wanted to keep up with the latest in the automotive field."

Other students agree, adding

Ken Meyer / *Rose Thorn*
Student Clint Hammes and GM Mentor Art McGrew work on a temporary motor.

that the experience is excellent. "The experience I am getting in advanced vehicle simulation and control algorithm development is unheard of for undergraduates," said junior mechanical engineer and control systems integration team leader, Matthew DeVries.

Students and advisors are not alone in the project. Aside from the school, they are supported by around 30 sponsors, including but not limited to the Department of Energy, General Motors (GM), who donated the Chevy Equinox, and National Instruments. The sponsors provide parts, funding,

and even training for students in cities such as Boston, Houston, and Detroit.

Challenge X teams follow GM's Global Vehicle Development process. The first year is for research and simulation. This year is the initial build phase - by this summer, the car will be taken to Phoenix for a test run. The third year is when the vehicle is refined.

Part of the challenge of the project is to successfully convert the Equinox into a viable hybrid in the three-year time frame. "A hybrid runs on more than one source of energy," explains Herniter. Teams select two or more of battery, electricity, hydrogen, solar power, hydraulics, and combustion power sources. For the combustion sources, teams are to use alternative fuels. Some fuels considered were the reformulated gas, E85, which is 85% ethanol and 15% gas, and B20, which is a mix of 20% bio-diesel and 80% diesel. "Our team is using electricity and bio-diesel combustion for power," said Herniter.

From a mechanical standpoint, teams have three methods of driving the car. One is to put the hybrid in series – that is, have the combustion charge the electric-

Ken Meyer / *Rose Thorn*
Neil Myers, Phillip Meiser (back), Trevor Acres, and Gareth Carlson work on the engine in the stripped-down Chevy Equinox.

ity – and drive the car from the battery. Another is to power the front from one source and power the rear from another.

According to Herniter, the type Rose has chosen to use is a split-power planetary gear set which is controlled by an electric supervisor controller. Through the use of computer simulation, a controlling scheme will be developed that will be exported to a chip to

be used in the real car.

Other issues include properly containing and controlling the power from the engines. The 336 volt battery and the presence of a high estimated 300 amp current means the vehicle has roughly 90 kW to manage.

What does the Challenge X program mean to Rose-Hulman?

Continued on page 3...

Winter 2005

New Australia telescope gathers light from dark skies

Richard Ditteon / Rose-Hulman Professor

Erin Hudson
Staff Cartoonist

On Friday, October 19, several Rose-Hulman students joined Professor Richard Ditteon in the demonstration of the capabilities of the new Oakley Southern Sky Observatory in New South Wales, Australia. Students, staff and faculty passing through the commons area in Moench hall were able to catch a glimpse of images of the sky as visible only from the southern hemisphere.

After nearly two years of planning and construction and approximately $200,000 invested, the 20-inch diameter telescope and its facilities are fully operational. Until now, students affiliated with astronomy at Rose-Hulman have been utilizing the eight telescopes located in the Oakley Observatory on the east side of campus, the largest of which measure 14 inches.

In addition to the increased aperture size, the new telescope offers the advantage of clear, dark skies due to the low light pollution and lack of cloud cover in Australia. Also, students may observe during the daytime, when it is dark in Australia; the telescope controls, as well as the opening and closing of the roof, can all be accessed remotely by Rose-Hulman faculty and students.

During Friday's demonstration, students of the Introduction to Astronomy class were able to slew the telescope and direct the camera to take images of several exotic objects, including planetary nebulae, star clusters, and galaxies. Also obtained were images of nebulae in the nearby Large and Small Magellanic Clouds, some of the closest neighboring galaxies to the Milky Way. These objects, as well as countless others, can be viewed using Rose-Hulman's facilites for the first time.

A grant from Terre Haute's Oakley Foundation helped Rose-Hulman build, equip, and operate an observatory in the Australian outback region. The new facility provides year-round astronomy study and research opportunities, and a southern hemisphere perspective to observations.

May 2006

Rose-Hulman's team won the Society of Automotive Engineers' collegiate Supermileage Competition with its fuel-efficient, one-person vehicle, achieving 1,541 miles per gallon to top 30 North American colleges and universities. Earlier, the team placed second in the Shell Eco-Marathon Americas Challenge at the California Speedway (Fontana, California), achieving 1,637 mpg with another motor configuration.

Rose-Hulman joined several other colleges in signing the American College & University President's Climate Commitment. The Institute formed a Sustainability Team that included students, faculty, and staff members. Rose's commitment included green cleaning, installing touchless soap and paper dispensers and energy-saving light bulbs, and making nighttime temperature checks in large areas.

ROBOTICS CERTIFICATE PROGRAM
PREPARES STUDENTS FOR TECHNOLOGY CHALLENGES

Faculty Advisor Matthew Boutell (middle), assistant professor of computer science and software engineering, helps freshman Corey Payne and Matt Runchey with a computer programming question in a course that's part of Rose-Hulman's new robotics initiative. Boutell is among four faculty members that are teaching courses in the program.

Elements of computer programming, technology, mathematics, science and engineering are coming together in a new robotics initiative at Rose-Hulman that could lead to students earning the college's first robotics certificate.

The new Multidisciplinary Educational Robotics Initiative (MERI) has been eagerly anticipated by students, faculty members and companies that are clamoring for graduates with robotics skills and programming knowledge. Matthew Boutell, assistant professor of computer science and software engineering, compares robotics to the popularity of the personal computer industry in the 1970s.

"It (robotics) is ready to explode. It is where technology is headed in the future, and a familiarity with robotics will give Rose-Hulman students and graduates an advantage in their careers," he says.

Robotics is a multidisciplinary field, blending mechanics, electronics, controls, and software, and requiring engineers to have deep enough knowledge where they can contribute within their specialty, but broad enough knowledge to understand other engineers. They must also be able to work in multidisciplinary teams. Rose-Hulman's MERI program, supported by a Faculty Success Grant

from the Lilly Endowment Inc., strives to meet all of these objectives, according to David Fisher (Mech. Eng., '00), assistant professor of mechanical engineering.

"Engineers in the future will need to understand and appreciate the components of the entire system at a high level to communicate with others, but tend to contribute primarily in a concentration (areas like kinematics, controls or computer programming)," he says.

So, rather than have a robotics major, Rose-Hulman's new robotics initiative will include students who are enrolled in academic majors that most interest them: mechanical engineering, electrical engineering, computer engineering, computer science or software engineering. They then earn the robotics certificate by taking seven courses covering a variety of areas, depending on major course of study. Some of the courses on the list cover such topics as mobile robotics, artificial intelligence, digital systems, wireless systems, mechatronic systems, image processing and microsensors. A senior-year multidisciplinary robotics project will cap the experience.

The robotics initiative started this fall with 18 freshmen and sophomore students enrolled in an Introduction to

Learning About Robotics: Freshman Ian Stevenson and Karl Heidtbrink work together to solve a problem on their Roomba robot during a recent competition in the Introduction to Robotics Programming pilot course being taught this fall.

Robotics Programming pilot course.

"Robotics links several disciplines together — computer programming, mechanics, electrical circuits and problem solving. It's a thrilling challenge that I like exploring," states Virginie Frizon, a freshman computer science major.

Rose-Hulman's MERI program hopes to help the college attract students who have robotics experience through the FIRST Robotics, Botball, First Lego League and other national competitions. These students are ready to learn more about programming, electronics, controls, artificial intelligence, robot vision and kinematics, according to Carlotta Berry, assistant professor of electrical and computer engineering.

Winter 2008

A new Multidisciplinary Educational Robotics Initiative was started to meet the rising student interests in this area and industry demand for graduates having robotics knowledge.

Rose-Hulman was among six U.S. colleges playing a role in developing next-generation lighting under a National Science Foundation Smart Lighting initiative. As an outreach partner, the Department of Physics and Optical Engineering received $500,000 over a five-year period to develop course materials to teach students about light-emitting diode (LED) technologies that could one day change the way the world is illuminated.

Winter 2008

May **2008**

The Human Powered Vehicle Team won its first
American Society of Mechanical Engineering's Human
Powered Vehicle Challenge, beginning a stretch of
capturing numerous titles.

Another series of projects over the course of three summers made a difference to the Dominican Republic community of Batey Cinco, including constructing a hurricane- and earthquake-resistant roof, providing a sanitation system for a former sugar plant, and installing a septic system for a community health clinic.

Engineers go batey

Alex Mullans
Sports Editor

Rose-Hulman's organization Engineers Without Borders (EWB) is currently exploring the possibility of a trip to the Dominican Republic to do some renovation and construction work. During the Thanksgiving break, three Rose-Hulman students (Daniel Giranda, Eric Hollenkamp, and Abby Grommet) went with Professor John Gardner and 1996 chemical engineering alumnus Levi Barclay to Batey Cinco Casas (in the Monte Plata region) to assess what work needed to be done to expand the medical clinic there.

For those unfamiliar with the region, a batey is like a company town in the United States built around a factory. Specifically, bateys were built around sugarcane plantations. In the profitable years, immigrants were brought from neighboring Haiti as workers. As business waned and the town began to die, the workers were left to fend for themselves as the plantations were shut down and the primary source of income removed.

Current statistics suggest that several hundred thousand people currently live in these semi-abandoned

John Gardner / Rose-Hulman

Students with the Rose-Hulman chapter of Engineers Without Borders assess the needs of a clinic during their recent trip to the Dominican Republic. In the proposed project, the students will renovate the facility to accomodate the growing medical needs of the surrounding community.

bateys. The non-profit Batey Relief Organization has set up a clinic in Cinco Casas where medical care is provided to anyone who needs it, regardless of their ability to pay.

The reason for the Engineers Withougt Borders trip seems simple: the clinic has seen a three hundred increase in need over the past two months and is need of more space to better serve their community. EWB was asked to help, and has already begun to do so: the Thanksgiving group tested water, assessed the structural stability of the existing buildings, and worked with the community to ensure that the proposed project would meet as many pressing needs as possible.

Rose-Hulman Engineers Without Borders hopes to complete the project in three years, a task that includes at least one more assessment trip as well as the actual construction period. Engineers Without Borders's president, Adam Kirchner (civil engineering major, Class of 2011), estimates that "construction will likely cost on the order of $250,000" and notes that the group's budget "relies heavily on donations and fundraising efforts." The club encourages people who may be interested in the project and others like it to contact the Rose-Hulman chapter of EWB and consider becoming a member.

Summer **2009**

February **2010**

Alumni and campus officials joined community, business, and government officials at a special Leading the Next Decade of Innovation Gala, conducted at the Indiana Roof Ballroom in downtown Indianapolis. Indiana Gov. Mitch Daniels was the first recipient of the Institute's Excellence in Innovation Award.

Fall 2010

THE ROSE THORN

Rose-Hulman Institute of Technology • Terre Haute, IN • HTTP//THORN.ROSE-HULMAN.EDU • Friday, September 10, 2010 • Volume 46 • Issue 1

Rose renovates Logan Library

Top, secured zone around Logan Library during contruction work. Bottom, proposed floor plan of the first floor of the newly renovated Logan Library.

Tim Ekl • *The Rose Thorn* **Stock Photo** • *Rose-Hulman*

Tim Ekl • *editor-in-chief*

Ed. note: an earlier version of this article ran in the freshman issue of The Rose Thorn. That article was written by Alex Mullans, editor-in-chief.

The renovation of Logan Library stretched into the first weeks of the 2010-2011 academic year as work crews rushed to complete the rebuild. Though originally slated to be complete by the start of the school year,

work on the main and upper floors of the library is now scheduled to last through September 20, according to Jan Jerrell, circulation coordinator.

While the entire book collection is being held in storage, library staff will attempt to retrieve books on request. In addition, inter-library loans are still available.

"We feel confident that these minor inconveniences will be worth the effort and that students, faculty and staff will be very pleased with

this significantly enhanced and functional space," Jerrell said in a campus-wide email.

The renovation project began in June and lasted the summer. New features to the reconstructed library are to include a café, additional individual and group workspaces, and extra seating for students.

"There are several reasons [why the library is being renovated]. The main one is to provide more learning space for students and the other

is to update the aesthetics of the library," said Rachel Crowley, Director of the Logan Library. She notes that "the Reference area and the [Digital Resource Center] will still be available for use, but in a more inviting atmosphere."

During the renovation, several student employees worked to place books in storage and, along with library staff, evaluate the existing collection and discard materials as necessary to make room for newer items.

A \$2.1 million renovation project of the Logan Library significantly increased student study and project spaces, consolidated the library collection, added a coffee bar, and improved the structural integrity of the campus landmark.

January
2011

Hall competition re-kindled with Greatest Floor competition

The Greatest Floor competition was started to help students living in residence halls to beat the winter "blahs" and be united by having floors working together to compete in 24 events over the course of 24 hours on a weekend. There's a "mystery" event at the end that makes each year's competition even more special and fun. Alumni now return to join in the festivities (and won the 2024 competition).

Speed 2 won the first ever Greatest Floor Competition, held ninth weekend of winter quarter by Student Affairs, by combining strong overall participation with wins at the Super Smash Bros. and basketball tournaments.

Liz Evans, a 2013 Math and Electrical Engineering graduate who also earned a Master of Science in Engineering Management in 2015, won the first of five high jump NCAA Division III national championships (2011 indoors and outdoors; 2012 indoors and outdoors; and 2013 outdoors). Evans was an eight-time All-American.

---

18 MAR 2011

SPORTS

7

Engineers successful at Indoor Nationals

Kurtis Zimmerman
sports editor

Liz Evans is #1

Sophomore Liz Evans wrapped up her indoor track and field season with a successful trip to the NCAA Division III National Championship. Clearing 5' 7 3/4" on her first attempt, Evans claimed the national championship in women's high jump, becoming the first female and fourth student-athlete in Engineer history to earn an individual national championship.

All smiles during her post-competition interview, Liz told the press "It's truly amazing to finally be a champion. I felt I may have given the title away the last two years, but I'm so happy that I did it this year."

Evans' #1 finish is the first individual national championship for an Engineer athlete since Matt Smith won the 100-yard breaststroke championship in 2003.

"It's exciting to rewrite the school record book in

Liz Evans became the first woman in Rose history to win an individual national championship
Linda Striggo • Striggo Photos

women's track and field," said Evans. And she has managed to do just that in both the high jump and long jump events. This marks Evans' third career All-American award, having finished runner-up in the high jump both indoors and outdoors last year. And she only has further ambition from here, telling the press on Saturday, "I would love to clear 5' 10" in the outdoor season."

Men's track qualifies

Three other Fightin' Engineer student-athletes also travelled to Capital University to compete in separate events.

Junior Sutton Coleman finished sixth in the 55-meter high hurdles with a time of 7.61 seconds after setting a school record with a 7.58-second run during qualifications on Friday. This marks his second career All-American award having finished eighth in the 400-meter intermediate hurdles at the 2009 Outdoor Championships.

Senior Derek Bischak claimed 11th place in the mile run with a time of 4:17.85, and junior Jeremiah Edwards rounded out the group with another 11th-place finish in the 55-meter dash.

This year marked the most Fightin' Engineer qualifiers since the 1985 Outdoor Championships, ending another successful indoor season.

Outdoor season begins

As the indoor track and field season wraps up, the outdoor season picks up again. In a preseason poll of

league coaches released earlier this week, the men's track and field team was unanimously selected to finish first among conference competitors. Such a finish would mark the fourth consecutive HCAC Outdoor Track and Field championship for the Engineers. The women's track and field team was voted to finish second closely behind Franklin College.

Returning for the men's team are Bischak and fellow senior Paul Bouagnon, who shared HCAC Athlete of the Year honors during the indoor season. Coleman earned the honors last year and will continue to be a contender during the outdoor season. Andrew Thompson was named HCAC Freshman of the Year and looks to compete in several outdoor events.

Competitors for the women's team are HCAC Field Athlete of the Year Liz Evans and fellow sophomore Tanya Colonna, competing in the pole vault event, and Creasy Clauer in the 800 meters. Erin Cox will likely compete in multiple events as well after earning HCAC Indoor Freshman of the Year honors earlier in the season.

The track and field team kicks its outdoor season with the Rose-Hulman Early Bird Invitational Saturday beginning at noon at the William Welch Track and Field Complex.

Winter **2011**

The Office of Global Programs was established to partner with students, faculty, staff, and the greater community to foster a campus environment that supports and promotes international understanding and engagement. The Institute has expanded study abroad with 21 countries overseas and increased spring and summer study courses and excursions that help teach students to be cognizant of global challenges and how to solve them in relation to their cultural, social, political, and geographic factors. It was renamed the Center for Global Engagement in 2018.

Spring 2011

NSBE News

VOLUME 4, ISSUE 1 — MAY 2, 2011

NSBE Small Chapter of the Year

Rose-Hulman chapter earns Regional and National recognition

This Spring twenty-three members of Rose-Hulman's National Society of Black Engineers (NSBE) chapter attended the NSBE National Convention in St. Louis, Missouri. The convention provided opportunities for these members to attend professional workshops, a two-day career fair with about 230 companies, a Graduate School fair, and to take practice GRCAT and GRE tests. Several awards were given out to recognize accomplished members, chapters, and regions. Members of the Rose-Hulman Chapter were pleasantly surprised when Rose-Hulman was nominated in both the NSBE Region IV and National Small Chapter of the Year.

The winners of these prestigious awards are chosen through extended Institutes for Chapter Development (ICD) reports which document efforts each chapter has made towards growth and improvement. For the Rose-Hulman chapter these efforts were seen through various programs. The chapter

The Rose-Hulman Chapter shows off the National Small Chapter of the Year Award

increased its visibility on campus through events like Soul Food Sunday and the annual NSBE Talent Show. The chapter also increased its significance in the Terre Haute community by tutoring at Chauncey Rose Middle School, which helped cultivate interests in math and science among middle school youth. Members of the Rose-Hulman chapter also participate in several campus-wide events like Intro for Tykes and the recent Adopt-Day tree planting. Others host multiethnic high school juniors who have been accepted to Rose-Hulman during the extremely successful Senior Weekend, which won recognition as Rose-Hulman's Best Student Event in 2011.

This year's goals were only achieved through outstanding leadership, efficient communication, and increased member involvement. The chapter hopes to build upon its achievements through new and improved programs concerning retention, academics and cultural awareness. This will help the chapter continue to fulfill the powerful NSBE Mission Statement, which is "To increase the number of culturally responsible Black engineers who excel academically, succeed professionally and positively impact the community."

Grace Johnson-Bunn, Chapter Secretary

The Rose-Hulman Chapter at the NSBE National Convention Golden Torch Awards

INSIDE THIS ISSUE

Books of the Engineer ... 2
NSBE Talent Show ... 2
Senior Weekend ... 3
Soul Food Sunday ... 3
Region IV Business Basics ... 4
Beauty Night Competition ... 5
Outgoing Presidents Remarks ... 5

ROSE-HULMAN INSTITUTE OF TECHNOLOGY CHAPTER

The National Society of Black Engineers chapter was selected as the Small Chapter of the Year after earning the Region 4 Chapter of the Year award, based on chapter leadership, membership growth, academic excellence, community service, and professional development activities.

The Student Innovation Center was opened to provide cooperative workspaces for student competition teams, club activities, and design projects. The 16,200-square-foot facility also had rooms for welding and painting. On September 22, 2012, the building was named Branam Innovation Center in memory of 1979 alumnus Matt Branam, who served as the Institute's 14th president.

October 2011

New buildings, furniture greet students on return to campus

Ranjana Chandramouli
staff writer

Rose-Hulman students this fall weren't returning to the same, old Rose that they were always used to. Couches, potted greens, tiled floors, light fixtures, and much more were replaced this summer as part of an aggressive and intensive renovation project led by the Facilities department here at Rose to update the campus as whole to keep up with the demands of the students. From the brand new Student Innovation Center (SIC) to the carpeted upper floors of Moench fall to ventless hoods in the Organic Chemistry lab to renovations in the Learning Center, the campus was bustling this summer with changes and innovations that changed the face of Rose-Hulman in less than 90 days.

The most noticeable renovations for most students' eyes are the ones in the main halls and common areas of the academic buildings. The upper floors of Moench are now carpeted and outfitted with the same style of desks and chairs that were introduced to Olin during its renovation last year. The Moench commons, by the mail rooms, is also outfitted with new tables, chairs, and power outlets and stripped of its delineating couches. However, many smaller renovations occurred throughout campus that are likely to go more unnoticed by the general student body. As part of the new 'green' initiative, more energy efficient lamps were installed in some classrooms through Olin and Crapo as well as through the main areas of the academic buildings. Aluminum handrails, a project started by Blue Key, were installed in Hatfield; the East stairwell in Deming was also updated – both of which were projects concerned with student safety. The organic chemistry lab was outfitted with new, ventless hoods and a new biology lab was created in the Myers building, while the Learning Center received its own revamp with new carpet and paint.

The starkest difference in the Rose-Hulman campus this fall, however, is the addition of a brand new Student Innovation Center, which is an addition to the current Facilities building north of Myers. While the renovations to Moench and the rest of the academic buildings fell in the school's capital budget for summer renovations, the creation of Student Innovation Center was motivated and implemented due to time pressure from the Indiana Department of Transportation regarding the South Campus. According to Michael Taylor, the Senior Director of the Facilities Department here at Rose-Hulman, INDOT planned to purchase the land; the need to vacate the premise and find new homes for such Rose clubs as SAE and EcoCar that were housed there "provided incentive to move judiciously and make alternative location arrangements [the building of the SIC] before the property had to be vacated."

However, this urgent need for space on campus blossomed into an inventive and modern place to house multiple teams on campus where there wasn't currently adequate room. Rose-Hulman's focus on innovation "helped drive," according to Taylor, "the discussion the development of a [shared space] for teams... such as Eco-Car, Formula 1 SAE Car,

The new Student Innovation Center is an extension of the Facilities Operations building on the north side of campus.
Rose-Hulman News

Team Rose Motorsports (TRM), Rose-Hulman Efficient Vehicles (RHEV), Concrete Canoe, Human-Powered Vehicle Team (HPVT) and ChemECar." A variety of difference clubs will be able to share space in the brand-new Student Innovation Center, providing them with a place on campus to meet or to work on their vehicles or projects.

Although already knee-deep in another project, Michael Taylor is, understandably, proud of the success of the entire renovation, especially considering the time crunch of a 90-day schedule from start to finish. Their goal was to have Rose-Hulman ready for students by the time they returned in the fall, making the remodeling process very aggressive and intense, made possible by "sound planning, a good construction team, and quality oversight and management... from the Institute itself," says Taylor. Like many other Rose-Hulman faculty and staff, Taylor believes in the success of the project because he believes that the new renovations will help the things that make Rose-Hulman truly new and fresh every year: the students. He remarks that "[the renovations made to Moench and the new Student Innovation Center] will impact [Rose-Hulman] students tremendously... and positively going forward.

The Office of Learning & Technology was established to provide instructional services and support faculty, students, and staff. The office offers educational planning and support services to facilitate the innovative and appropriate use of educational technology for face-to-face, hybrid, and online learning — elements that would be crucial in meeting the needs of providing online classes in response to the COVID-19 pandemic in the spring of 2020. The office was established following a generous gift from 1984 alumnus and emeritus trustee Mike Hatfield.

2012

April 3, **2012**

Six Rose professors earn national distinction

Ranjana Chandramouli
staff writer

Six of Rose-Hulman's many gifted faculty members were selected among the 300 best in the country, according to a book published by The Princeton Review last week. Honorees (pictured clockwise, from top) include Dr. Bill Weiner, of the Applied Biology and Biomedical Engineering department, Dr. Richard Stamper and Dr. Phillip Cornwell, both professors of Mechanical Engineering, and Dr. Yosi Shibberu, Dr. Elton Graves and Dr. Diane Evans, who are all members of the Mathematics department here

at Rose-Hulman.

An intensive process that collected data from thousands of colleges across the country helped The Princeton Review and RateMyProfessors. com compile a list of the nation's best. Starting with surveys and online data, the two collaborating companies were able to pinpoint colleges "at which students highly rated their professors' teaching ability and accessibility," according to The Princeton Review Press Release. From these colleges, they were able to establish a base list of 1,000 professors, which were then whittled down using input from college administra-

tors and students and surveys of the professors themselves. The final result represents the 'best' of the over 1.8 million college professors across the country, including six of Rose-Hulman's own faculty members.

Rose-Hulman was the only university from Indiana on the list of 122 colleges in America, and the six professors from Rose represented two of the nine engineering professors and three of the 32 math professors on the list. In honor of these professors as well as the entire Rose-Hulman faculty, a banner was set out in the Union earlier this week to commemorate and thank those who

give so much to make Rose-Hulman among the best in the nation. What sets Rose-Hulman faculty apart is their dedication to their students and their commitment to a personalized learning experience.

Matt Branam, our university president, proudly remarked, "Our professors have dedicated their professional lives to the success of our students. I know they would appreciate knowing how they have made a difference in yours."

A complete list of the professors is available online on the Princeton Review website: http://princetonreview.com/best-professors.

Rose-Hulman News

Six faculty members were featured among America's best 300 undergraduate college professors by The Princeton Review for a new book highlighting excellence in teaching. Professors profiled were Phillip Cornwell, PhD, mechanical engineering; Diane Evans, PhD, mathematics; Elton Graves, DA, mathematics; Yosi Shibberu, PhD, mathematics; Richard Stamper, PhD, mechanical engineering; and William Weiner, PhD, biology and biomedical engineering. Rose-Hulman was the only Indiana college or university to have a professor selected to the prestigious list, and only 10 other institutions had six or more professors chosen.

August 2012

Lakeside Hall's first residents pleased with results

Ranjana Chandramouli
staff writer

The first set of Lakeside Hall residents moved in before the start of the school year. The new residence hall received positive feedback. Several residents were gracious enough to give us their personal opinions of the living quarters.

"The rooms are really nice, ...like I am living in a hotel. I am able to look down either side of the hall just by stepping out of my room. The lobby is nice because it allows people to hang out and be a little loud outside of their room. The one big change I think all of us RA's here would want to make would be to the doors. The doors are like hotel doors, as in they shut automatically. This is bad for the open-door policy we like to create on the floors, but all the residents are working around it, and we got door stops to help keep them open."

– Ashley Kohls, Senior ME, RA on 3rd floor

"I like the way that the rooms are set up. With the bedrooms on one side of the apartment, it becomes more private and you are less disturbed when you are sleeping. I like the dark finish of the furniture. The new apartments will be a good living space for this year and years to come for students."

– Jaci Dalton, Junior EE

"I love the fact that Lakeside Hall is doing its part to conserve energy resources. The biggest thing I would change would be to have an aluminum and paper/cardboard recycling bin along with our trash cans in our rooms so that it is much easier to reduce waste."

– Chelsea Copenhaver, Junior BE

Lakeside residence hall on the west edge of campus was opened to house 240 students in the four-floor, 75,000-square-foot building to meet the growing demand for campus housing. It was also the first LEED Silver certified building on campus, reflecting the college's commitment to sustainability.

The first "living laboratory," the 1,350-square-foot William Alfred Cook Laboratory for Bioscience Research, was opened on the south side of Crapo Hall to showcase the increasing role of the life sciences on campus. A $500,000 gift from Carl Cook provided funding in honor of his late father, life sciences pioneer William Alfred Cook, who was a longtime advocate for the Institute's biosciences programs. Carl Cook became Board of Trustees Chair in 2023.

September
2012

William Cook Bioscience Laboratory dedicated

Members of the Rose-Hulman community gathered Thursday for the dedication of the William Cook Laboratory for Bioscience Research.

Pictured at the ribbon-cutting ceremony from left to right are Dr. Peter Coppinger, senior Allie Williams, President Rob Coons, William Cook's son Chad Cook, and Chairman of the Board of Trustees William Fenoglio.

Rose-Hulman goes green with new lab
Greenhouse expands hands-on research in biology department

The biosciences laboratory will allow students unique opportunity to examine living and isolated specimens.
Rose-Hulman News

Katie Dial • staff writer

Rose-Hulman received a $500,000 gift from Carl Cook, on behalf of his father, William Alfred Cook. William Alfred Cook supported the growth of the biosciences at Rose-Hulman, and is contributing posthumously towards the William Alfred Cook Laboratory for Bioscience Research: a 1,350 square feet greenhouse that will be nestled between Crapo, Logan Library and the Root Quandrangle

A majority of Rose students end up taking a biology class before they leave, and everyone who takes a biology lab will benefit from the addition of the Cook Laboratory. Dr. Michael Mueller, head of the Chemistry department, said "I think that the number of students that want to take biology has certainly been growing. The greenhouse is a tool. It's like any other laboratory instrument or device. One of the things that really makes a Rose education great is that so much of what we do is hands-on. And what is it that we expect to come from it being hands-on? We don't really expect everyone to master every skill they come in contact with, but they sort of grow an appreciation for what's involved in different processes. It's important for Rose students to understand what it takes to grow things--what it takes to isolate living substances.

Continued on page 3.

FIRST Robotics' new Crossroads Regional brought more than 4,000 people associated with 50 teams to the Sports and Recreation Center to measure the effectiveness of their robots and test the power of collaboration during the regional round of the competition. FIRST Robotics founder Dean Kamen was a surprise visitor during the event. A second regional was hosted in March 2014, and the FIRST Indiana Robotics Championship was hosted in April 2022.

October **2013**

The 20th anniversary of the partnership with the Kanazawa Institute of Technology was celebrated by hosting a 19-member delegation from KIT for educational workshops and social events. The visit was marked by planting a grove of 40 Japanese Sakura trees near the White Chapel.

"Ideas Worth Spreading" was the topic of the first TEDxRoseHulman, an independent student-organized TED event conducted in Hatfield Hall. Guest speakers, faculty, and students presented a variety of topics that explored creative worlds and provided students with perspectives about STEM career options. Other TEDx RHIT events followed in 2017 and 2024.

Spring **2014**

An expanded 1,800-square-foot laboratory for Micro-Nano Device and Systems (MiNDS) on the first floor of Myers Hall has allowed students access to state-of-the-art equipment to complete a variety of micro and nano technology-related projects. Course work has spanned the fields of material science, chemical detection, optics, power generation, and bio-MEMS.

Continuing to evolve, the Department of Applied Biology and Biomedical Engineering was renamed to the Department of Biology and Biomedical Engineering, offering a high-quality, hands-on educational experience with award-winning professors.

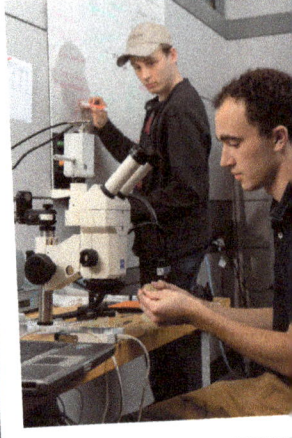

April **2014**

Summer **2014**

Rose-Hulman Ventures under-
went a transformation on the
Institute's South Campus site,
with the conversion of 16 small
spaces into five major engineering labs; the addition of a heavy
fabrication space; increased
lab space for projects needing
special safety, security, or other
isolation requirements; and the
creation of an open software
development lab within the
development server hardware.

A new academic major in biomathematics was offered to provide students the analytical tools necessary to use applied math in support of the life sciences. The area blends applied mathematics, the fundamentals of biology, and computational analysis. The program introduces students to the fields of computational biology, mathematical biology, bioinformatics, systems biology, and biostatistics.

September **2014**

A $1 million Lilly Endowment Inc. grant established the ESCALATE program, a living and learning community that provides students with the essential principles of entrepreneurial and business success, and allows them to take upper-level technical entrepreneurship courses and become involved in co-curricular activities, such as the student entrepreneurial club. An Entrepreneurial Intern Program was enhanced by increasing collaborations with entrepreneurs across Indiana and partnerships with alumni, local communities, Indiana businesses, and other educational institutions.

Rose-Hulman received the 2015 Best In Class award among Kern Entrepreneurial Engineering Network (KEEN) institutions for producing graduates with an "entrepreneurial mindset" characterized by exercising curiosity, seeking connections, and creating value. A $2.25 million grant in 2014 from the Kern Family Foundation encouraged further development of entrepreneurially-minded learning programs through educational practice, faculty engagement, and student experiences on campus.

January 6, 2016

KEEN
KERN ENTREPRENEURIAL
ENGINEERING NETWORK

March **2017**

A one-of-its-kind McLaren P1 hypercar owned by 1969 alumnus G. Felda Hardymon brought in a world record $2.39 million price in an auction that endowed the Alfred R. Schmidt Chair for Excellence in Teaching. The faculty chair supports creation of transformative learning experiences in the classroom on campus and provides opportunities for national engagement with other exceptional scholars. A loyal 1949 alumnus, Schmidt, was one of Rose's longest-serving faculty members, with 46 years as a mathematics professor.

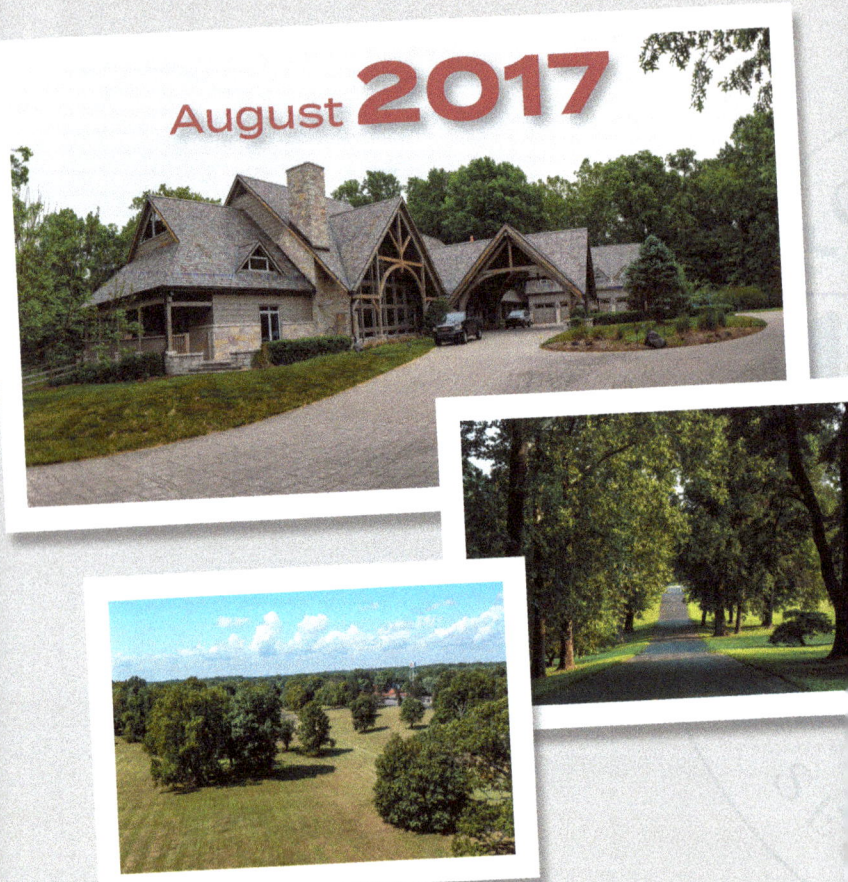

August 2017

Rose-Hulman's footprint expanded with the acquisition of approximately 1,100 acres of land on the Hulman Farm property just south of campus. The property includes a large residence as well as a historic family lodge, several outbuildings, and acres of farmland and forested terrain.

A newly expanded and renovated student union was opened with nearly 104,000 square feet of space where the campus community could come together, relax, and enjoy healthy dining. Also added were new outdoor patio and gathering spaces, a 120-person dining/meeting area, new spaces for career services and diversity and inclusion offices, and new/refurbished conference rooms. The project also expanded green spaces on campus with a pond near Baur-Sames-Bogart Hall (affectionately known as Scum Pond) being filled in. The project was supported by Linda and Mike Mussallem, a 1974 alumnus and trustee. It cost $25 million.

A glass-enclosed Pi-Vilion (named by students) was added in 2019, also supported by the Mussallems.

May 2018

June 2018

A new major in engineering design was added to provide students with a broad foundation of engineering skills and a deep, multi-disciplinary experience in design to meet the needs of a dynamic innovation economy. The unique program awards an academic degree to students completing a comprehensive four-year design-centric curriculum, along with a traditional engineering concentration. The program earned accreditation status in 2023 from ABET's Engineering Accreditation Commission.

The Richard J. and Shirley J. Kremer Innovation Center was opened to provide additional hands-on design and laboratory opportunities for students. The 13,800-square-foot facility, located adjacent to the Branam Innovation Center, has high-tech tools and testing equipment, project and robotics workstations, classrooms, and a conference room for students and faculty. The building is named for the 1958 alumnus and his wife.

April **2019**

ROSE·HULMAN
INSTITUTE OF TECHNOLOGY

**Kremer Innovation Center
"The KIC"**

Named in recognition of a generous gift from Richard J. (CHE, 1958) and Shirley J. Kremer.

The KIC is designed to provide space for hands-on learning and academic exploration. Throughout their lives, the Kremers have exemplified hard work and determination, and the innovative and entrepreneurial spirit that they hope to inspire in others.

Dedicated April 3, 2019.

With the Department of Humanities and Social Sciences offering
nearly 20 courses in the arts, "the Arts" was added
to the department's name — becoming the
Department of Humanities,
Social Sciences, and the Arts.

Fall **2019**

DATA SCIENCE

Today, the world is awash in trillions of gigabytes of data, opening the door wide for those who can discover patterns, build algorithms, and unlock the hidden stories in vast reams of data. Here's why studying data science at Rose-Hulman is the perfect choice:

- It blends the fields of mathematics, statistics, and computer science to give you the tools you'll need to succeed as a professional data scientist.
- You'll work closely with incredible faculty getting in-depth experience in data engineering, data analysis and data intelligence.
- You'll be exposed to machine learning, data engineering, and artificial intelligence.

Above all, Rose is a great place to be yourself, learn and grow. You'll love our beautiful campus, student activities and career services, and our sterling reputation for academic excellence will remain with you throughout your life. Your education will stretch your limits while giving you the tools to make the world a better place.

Got questions? Contact
Professor Sriram Mohan, department head
mohan@rose-hulman.edu • 812-877-8819

October 2019

Data science was added as a secondary academic major, along with a multidisciplinary minor that combines science, mathematical algorithms, and computer science principles to extract knowledge and insights from data. The new major provides students with in-depth hands-on experience in data engineering, data analysis, machine learning, and artificial intelligence.

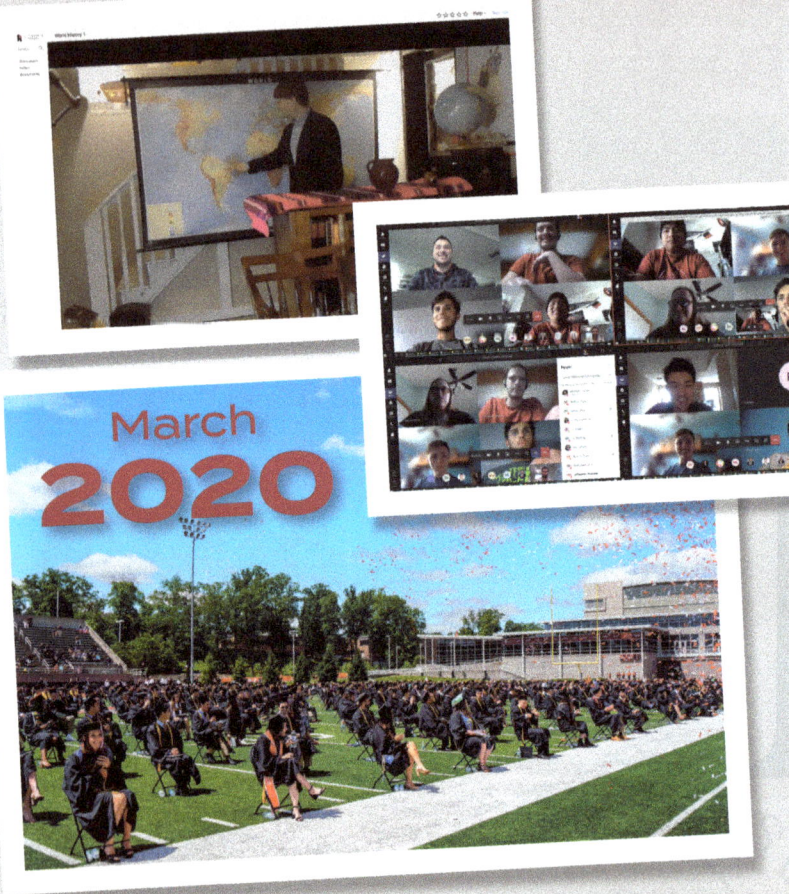

March
2020

Courses were taught online exclusively for the first time, as campus operations were shut down by the COVID-19 pandemic. Other adjustments were made, including an online Commencement ceremony. The pandemic continued into 2021, and the Commencement ceremony was held outdoors on Phil Brown Field at Cook Stadium.

A $10 million gift from alumnus and then-Board Chair Niles Noblitt and his wife, Nancy, supported a first-of-its kind college-based merit and leadership Noblitt Scholars program designed to transform the world's most gifted STEM students into future science and industry leaders. The program, started for the 2021–22 school year, takes to a new level the individual attention and support that has helped make Rose-Hulman one of the premier STEM colleges in the nation. Besides receiving scholarships, approximately 35 annual scholars also have the opportunity to participate in educational travel, mentoring, and leadership development workshops.

Noblitt
SCHOLARS PROGRAM
ROSE-HULMAN INSTITUTE OF TECHNOLOGY

August
2020

November 18, **2020**

A new Rose Squared program was offered, combining a bachelor's and master's degree pairing to allow students to use eligible credits earned before attending Rose to obtain two degrees in just four years at no additional cost. The program was launched with a Master of Engineering Management and has continued to expand to other academic departments.

The *MIND*ful College Connections not-for-profit consortium was established to help Rose-Hulman, DePauw University, and Saint Mary-of-the-Woods College more efficiently and effectively focus on preventive care strategies. The consortium also aims to expand student access to mental health services—such as psychiatrists, counselors, and telehealth services—while allowing each institution to tailor resources to their individual campus needs. This coalition is supported by an $8.1 million grant from Lilly Endowment Inc.

The Mission Driven comprehensive fundraising campaign was completed with more than $250 million being raised to increase student scholarships ($127 million), support new learning spaces to campus ($30 million), and solidify Rose-Hulman as a national leader in undergraduate STEM education ($93 million). The campaign raised $50 million over its last year and more than $100 million since 2018. "The Mission Driven Campaign for Rose-Hulman has transformed our campus and prepared us to better meet the challenges of being a top STEM college focused on undergraduate education," said President Robert A. Coons.

August **2021**

A new 70,000-square-foot academic building was opened with its state-of-the-art design studios, collaborative workspaces and science laboratories, and flexible classrooms. The central atrium provides a "window into Rose" with interior and exterior glass that showcases all the work taking place within the building. A large-scale Depth of Field artwork, underwritten by alumnus/trustee Tom Dinkel and his wife, Susie, is a focal point of the atrium. The building later became Indiana's first building to earn full WELL Certification for design and technology.

March **2022**

The Wabash Battalion Army ROTC unit earned the MacArthur Award as the top program in the military branch's 7th Brigade. Later, the battalion received the prestigious Department of Defense ROTC and Partner Institution Excellence Award in recognition of being the top performing collegiate program among all military branches for the 2021–22 school year.

Completion of a four-phased extensive renovation of Moench Hall created a new Moench Café while updating faculty office areas, laboratories, and class-rooms. The renovations provided new office spaces, improved accessibility, added student study spaces, and improved energy efficiencies.

September **2023**

A groundbreaking ceremony marked the start of construction on a new $30 million, four-floor residence hall for 160 first-year students, scheduled to be open by the Fall 2025. The project will allow the Institute to make substantive renovations to Speed Hall during the 2025–26 school year.

September 2023

The 100th celebration of the Institute's Homecoming bonfire tradition surprised and dazzled alumni, family members, students, and other guests with a spectacular drone show that embodied the event's special significance. Five hundred drones lit up the skies with images of Rose-Hulman's familiar RH logo, Rosie mascot, a variety of STEM academic objects, an outhouse, and a cannon (whose blast kicked off the bonfire and fireworks show). The show was designed by Texas-based Sky Elements, co-founded by 2016 mechanical engineering alumnus Tyler Johnson.

Development began for an entrepreneurial ecosystem called Innovation Grove, which will feature the relocation of Rose-Hulman Ventures and the addition of other new amenities, including a partnership with Union Health and Indiana Joint Replacement Institute, to provide students with additional internships and hands-on experiences to gain the skills needed to thrive in a rapidly changing world. In August 2024, Rose-Hulman was awarded its largest grant in Institute history — $30.5 million — by Lilly Endowment Inc. in support of its "Trails to Innovation" initiative as part of the Innovation Grove project.

March 2024

A special campus Sesquicentennial Celebration on April 4 had students, faculty, and staff commemorating Rose-Hulman's 150-year legacy as a leader of undergraduate science, engineering, and mathematics, while also ushering in the next chapter of the Institute's future through the launch of the "Advancing by Design" Strategic Plan.

Institute Presidents

1. **Charles O. Thompson**: March 7, 1883 – March 17, 1885 (died)
   Clarence Abiathar Waldo (acting): 1885 – 1886

2. **Thomas Corwin Mendenhall**: September 1886 – July 1889
   Clarence Abiathar Waldo (acting): September 1889 – June 1890
   Carl Leo Mees (acting): September 1890 – December 1890

3. **Henry Turner Eddy**: January 1891 – September 1894
   Carl Leo Mees (acting): 1894 – 1895 (school year)

4. **Carl Leo Mees**: September 1895 – 1919 (took a leave of absence 1916 – 1917)
   John White (acting): 1916 – 1917, 1919 – 1921

5. **Philip B Woodworth**: May 1921 – May 1923 (indefinite leave of absence)
   Frank C. Wagner (acting): May 1923 – Fall 1924

6. **Frank C. Wagner**: Fall 1924 – November 21, 1928 (died)
   John B. Peddle (acting): November 24, 1928 – September 1, 1930

7. **Donald B. Prentice**: February 1, 1931 – June 1948
   Carl Wischmeyer (acting): June 1948 – January 1949 (resigned)

8. **Ford L. Wilkinson**: January 1949 – September 1958 (died)
   Herman A. Moench (acting): 1958 – 1959 school year

9. **Ralph A. Morgen**: September 1, 1959 – August 31, 1961
   Herman A. Moench (acting): September 1961 – August 1962

10. **John A. Logan**: September 1, 1962 – August 31, 1976

11. **Samuel F. Hulbert**: September 1, 1976 – July 1, 2004

12. **John J. Midgley**: July 1, 2004 – June 2005
    Robert Bright (interim [Chair of the Board of Directors]):
    July 1, 2005 – July 1, 2006

13. **Gerald S. Jakubowski**: July 1, 2006 – July 1, 2009
    Matt Branam (interim): July 1, 2009 – December 3, 2009

14. **Everett Matthew Branam**: December 4, 2009 – April 20, 2012 (died)
    Robert A. Coons (interim): April 21, 2012 – April 30, 2013

15. **James C. Conwell**: May 1, 2013 – November 15, 2018

16. **Robert A. Coons**: November 15, 2018 – Present

2024
Rose-Hulman Statistics

- Total enrollment: 2,300+

- Students annually represent nearly all 50 states and approximately 30 countries outside the U.S.

- 99% placement rate average for recent graduating classes

- Average starting salary is approximately $80,000 for recent graduates

- 90+ student clubs, organizations, and competition teams

- Nearly one in four students are student-athletes on one or more of Rose-Hulman's 18 NCAA Division III teams

Undergraduate Degree Programs

- Biochemistry
- Biology
- Biomathematics
- Biomedical Engineering
- Chemical Engineering
- Chemistry
- Civil Engineering
- Computer Engineering
- Computer Science
- Electrical Engineering
- Engineering Design
- International Computer Science
- Mathematics
- Mechanical Engineering
- NanoEngineering
- Optical Engineering
- Physics
- Software Engineering

* *Second majors only are offered in Computational Science, Data Science, International Studies, and Molecular Biology*

** *Graduate programs of study are offered in Biomedical Engineering, Chemical Engineering, Chemistry, Civil and Environmental Engineering, Electrical and Computer Engineering, Engineering Management, Mechanical Engineering, and Optical Engineering.*

A Message from President
Robert A. Coons

As we reflect on the illustrious history of Rose-Hulman Institute of Technology, we are reminded of the remarkable journey that began in 1874. Our founders, led by Chauncey Rose and guided by our first president, Charles O. Thompson, envisioned an institution that would stand the test of time, dedicated to the pursuit of knowledge and the betterment of  society. Over the past 150 years, Rose-Hulman has not only realized this vision but has exceeded it in countless ways.

Our present success is built on the foundation laid by the pioneers who came before us. From the contributions of Herman Moench in the classroom to the transformative philanthropy of Anton Hulman and the strategic leadership of Sam Hulbert, every era has brought new achievements and milestones. Today, we are proud to be recognized as a leader in undergraduate engineering education, a testament to the hard work and dedication of our entire community.

Looking ahead, we are excited about the future and the opportunities that lie before us. Our new strategic plan, "Advancing by Design," launched this year in tandem with our sesquicentennial celebrations, sets a bold vision for the next chapter of Rose-Hulman. This plan will guide us as we continue to innovate, inspire, and lead in our fields of study, ensuring that we remain at the forefront of technological education.

I invite all Rose-Hulman alumni, donors, and friends to join us on this journey. Your time, talents, and treasure are invaluable as we strive to build on our legacy of excellence. Whether you engage with us through mentoring students, participating in events, or philanthropically supporting our initiatives, your involvement makes a difference.

- For more information on how to give back to Rose-Hulman, please visit rose-hulman.edu/give

- To share your stories, read about others, and find sesquicentennial events, visit rose-hulman.edu/150

- To learn more about our strategic plan, visit rose-hulman.edu/strategicplan

- Keep up with the latest alumni news and campus events, update your contact information, and find other alumni resources at rose-hulman.edu/alumni

Thank you for being a part of the Rose-Hulman community and for your continued support as we look forward to an even brighter future.

Robert A. Coons
President